Building Generative AI Agents

Using LangGraph, AutoGen, and CrewAI

Tom Taulli
Gaurav Deshmukh

Apress®

Building Generative AI Agents: Using LangGraph, AutoGen, and CrewAI

Tom Taulli
Monrovia, CA, USA

Gaurav Deshmukh
Tarzana, CA, USA

ISBN-13 (pbk): 979-8-8688-1133-3
https://doi.org/10.1007/979-8-8688-1134-0

ISBN-13 (electronic): 979-8-8688-1134-0

Managing Director, Apress Media LLC: Welmoed Spahr
Acquisitions Editor: Shiva Ramachandran
Development Editor: James Markham
Project Manager: Jessica Vakili

Distributed to the book trade worldwide by Springer Science+Business Media New York, 1 New York Plaza, New York, NY 10004. Phone 1-800-SPRINGER, fax (201) 348-4505, e-mail orders-ny@springer-sbm.com, or visit www.springeronline.com. Apress Media, LLC is a Delaware LLC and the sole member (owner) is Springer Science + Business Media Finance Inc (SSBM Finance Inc). SSBM Finance Inc is a **Delaware** corporation.

For information on translations, please e-mail booktranslations@springernature.com; for reprint, paperback, or audio rights, please e-mail bookpermissions@springernature.com.

Apress titles may be purchased in bulk for academic, corporate, or promotional use. eBook versions and licenses are also available for most titles. For more information, reference our Print and eBook Bulk Sales web page at http://www.apress.com/bulk-sales.

If disposing of this product, please recycle the paper

Table of Contents

About the Authors

Tom Taulli is a consultant to various companies, such as Aisera, a venture-backed generative AI startup. He has written several books like *Artificial Intelligence Basics* and *Generative AI*. Tom has also taught IT courses for UCLA, PluralSight, and O'Reilly Media. For these, he has provided lessons in using Python to create deep learning and machine learning models. He has also taught on topics like NLP (natural language processing).

Gaurav Deshmukh is a highly skilled technology leader with over a decade of experience driving transformative software engineering initiatives. Throughout his career, he has held pivotal technical roles at prominent companies such as Guidewire, Cigna, Home Depot, American Agricultural Laboratory (AmAgLab), Tata Elxsi, and Amdocs. Gaurav's expertise encompasses a range of cutting-edge technologies, including cloud computing, cybersecurity, software automation, data engineering, and full-stack development with various programming languages and web technology frameworks. He employs his vast knowledge to create innovative solutions that optimize workflows and drive business growth. Gaurav holds both an MBA and a master's degree in Computer Science, with a focus on data warehousing and computer vision. He is dedicated to elevating the strategic role of software engineering in delivering business value.

CHAPTER 1

Introduction to AI Agents

Andrew Ng is a towering figure in the AI world. He has the rare blend of being an academic and entrepreneur.

When many in the tech world were focused on the dot-com boom during the 1990s, Ng saw AI as more interesting. While at Bell Labs, he worked on evaluating models, improving feature selection, and using reinforcement learning.

He would go on to get his master's degree in Electrical Engineering and Computer Science from the Massachusetts Institute of Technology (MIT) and a Ph.D. in Computer Science from the University of California, Berkeley. His thesis was about reinforcement learning.

Ng would become a professor at Stanford. His course, which was CS229, was the most popular among students. He was also one of the first to see the usefulness of GPUs (Graphics Processing Units) for AI systems.

Ng would eventually apply his AI skills to the business world. He became the chief scientist at Baidu and helped to create Google Brain.

Then in 2011, he led the development of Stanford's MOOC (Massive Open Online Courses) platform. It would quickly attract large numbers of students.

© Tom Taulli, Gaurav Deshmukh 2025
T. Taulli and G. Deshmukh, *Building Generative AI Agents*,
https://doi.org/10.1007/979-8-8688-1134-0_1

Ng leveraged this experience by cofounding Coursera, which is one of the world's top online learning platforms. The company went public in 2021, with a market value of nearly $6 billion. Currently, it has about 148 million registered users and has partnerships with more than 325 universities and companies.[1]

After this, Ng founded other companies like DeepLearning.AI and Landing AI. He even has launched a venture capital fund.

No doubt, Ng has a knack for understanding trends—especially in the field of AI. This is someone who you should not bet against.

Then what is he looking at next? Where does he see the biggest opportunities?

It's with AI agents. He has noted that they are an "exciting trend" and something you "should pay attention to."[2] He has also said:

> *AGI (Artificial General Intelligence) feels like a journey rather than a destination. But I think … agent workflows could help us take a small step forward on this very long journey.*[3]

Ng is far from an outlier. Many of tech's most influential people are optimistic about AI agents.

Just look at Bill Gates. In his blog, he wrote:

> *In the computing industry, we talk about platforms—the technologies that apps and services are built on. Android, iOS, and Windows are all platforms. Agents will be the next platform.*

In his post, he details how software has changed little since he started Microsoft during the mid-1970s. The applications are "pretty dumb."

---

[1] https://investor.coursera.com/overview/default.aspx
[2] https://www.youtube.com/watch?v=sal78ACtGTc&t=125s
[3] https://www.youtube.com/watch?v=sal78ACtGTc&t=125s

But AI agents will change everything. A key part of this will be due to a system's understanding of your "work, personal life, interests, and relationships." In other words, software will become very smart—and much more useful and productive.

According to Gates:

> *Imagine that you want to plan a trip. A travel bot will identify hotels that fit your budget. An agent will know what time of year you'll be traveling and, based on its knowledge about whether you always try a new destination or like to return to the same place repeatedly, it will be able to suggest locations. When asked, it will recommend things to do based on your interests and propensity for adventure, and it will book reservations at the types of restaurants you would enjoy. If you want this kind of deeply personalized planning today, you need to pay a travel agent and spend time telling them what you want.*[4]

Then there is this take from McKinsey, which is one of the leaders in helping companies leverage AI technologies:

> *The value that agents can unlock comes from their potential to automate a long tail of complex use cases characterized by highly variable inputs and outputs—use cases that have historically been difficult to address in a cost- or time-efficient manner. Something as simple as a business trip, for example, can involve numerous possible itineraries encompassing different airlines and flights, not to mention hotel rewards programs, restaurant reservations, and off-hours activities, all of which must be handled across different online platforms. While there have been efforts to automate parts of this process,*

---

[4] https://www.gatesnotes.com/AI-agents

much of it still must be done manually. This is in large part because the wide variation in potential inputs and outputs makes the process too complicated, costly, or time-intensive to automate.[5]

Note   Sonya Huang is a partner at Sequoia Capital. She has backed some of the hottest generative AI startups like Hugging Face, Glean, and LangChain.[6] According to her: "One of our core beliefs is that agents are the next big wave of AI, and that we're moving as an industry from copilots to agents."[7]

What Are AI Agents?

There is no clear-cut definition of AI agents. But this should come as no surprise. The category for AI agents is still in the nascent stages—and the technology is moving quickly. Just as the Internet grew to encompass a vast array of applications and services, AI agents are likely to undergo a similar trajectory of rapid development and diversification. This means that developers are at a point where significant opportunities for growth and excitement abound.

Yet we still need a basic definition. So what should this be? A good place to start is with one of the pioneers of the generative AI revolution, Harrison Chase. He is the cofounder of LangChain, which is one of the most popular development frameworks for this technology.

---

[5] https://www.mckinsey.com/capabilities/mckinsey-digital/our-insights/why-agents-are-the-next-frontier-of-generative-ai

[6] https://www.linkedin.com/in/sonyaruihuang/details/experience/

[7] https://www.sequoiacap.com/podcast/training-data-harrison-chase/

Here's how he defines generative AI agents:

The way that I think about agents is that it's when an LLM is kind of like deciding the control flow of an application. So what I mean by that is if you have a more traditional kind of like RAG chain, or retrieval augmented generation chain, the steps are generally known ahead of time, first, you're going to maybe generate a search query, then you're going to retrieve some documents, then you're going to generate an answer. And you're going to return that to a user. And it's a very fixed sequence of events.[8]

And I think when I think about things that start to get agentic, it's when you put an LLM at the center of it and let it decide what exactly it's going to do. So maybe sometimes it will look up a search query. Other times, it might not, it might just respond directly to the user. Maybe it will look up a search query, get the results, look up another search query, look up two more search queries and then respond. And so you kind of have the LLM deciding the control flow.

Another way to look at AI agents is to understand their components. They include reflection, tools, memory, planning, multi-agent collaboration, and autonomy.

Let's take a look at each.

Reflection

Reflection in AI agents refers to the ability of a system to inspect and adjust its own cognitive processes. This self-awareness allows the AI to scrutinize its decision-making, learning patterns, and problem-solving approaches. By engaging in reflection, AI can break down intricate challenges, extract insights from its experiences, and offer clearer justifications for its conclusions.

---

[8] https://www.sequoiacap.com/podcast/training-data-harrison-chase/

Recent research, such as the Reflexion framework, has demonstrated the significance of self-reflection in enhancing AI capabilities. Reflexion uses verbal self-reflections to generate valuable feedback for future trials, storing this feedback in the agent's memory. This process involves an iterative optimization where the agent evaluates its actions, receives feedback, and adjusts its behavior accordingly. This method has shown improvements in tasks like decision-making, reasoning, and programming.

This metacognitive ability enhances AI systems' flexibility and resilience. As the AI evaluates its past performance and outcomes, it can refine its strategies and expand its autonomous capabilities. The process facilitates error detection, strategic evolution, and more efficient goal attainment. For example, Reflexion agents have demonstrated improved performance in environments such as AlfWorld and on tasks like search-based question answering and code generation.[9]

Tools

Tool use in generative AI agents refers to their ability to interact with external tools, APIs, or software to enhance their capabilities and perform complex tasks. This feature allows AI systems to go beyond their core functions like language or image generation. AI agents can access up-to-date information, retrieve real-time data, perform calculations, manipulate files, and automate workflows by chaining multiple actions together. This integration significantly improves their accuracy and expands their domain knowledge.

Examples of tool use in generative AI include web browsing for current information, code execution in real-time environments, data analysis and visualization, calendar management, file operations, and complex

---

[9]https://ar5iv.labs.arxiv.org/html/2303.11366

mathematical computations. These capabilities enable AI agents to handle a broader range of tasks effectively. For instance, Salesforce's Einstein GPT integrates with CRM tools to provide AI-generated content across various business functions. Similarly, AWS's Solution Architect Agent uses custom-built tools to query AWS documentation, generate code, and create architectural diagrams.

Complex tasks often occur in dynamic environments where solutions are not immediately apparent or may change unexpectedly. For instance, data sources might be unavailable, requiring the search of alternative sources, or actions might have unforeseen side effects. In app development, an initial API request might fail due to network issues, incorrect argument formats, or changes to the API itself. To adapt, an agent might need to retry the request with different parameters based on feedback, such as error messages, or seek explicit human assistance.

What if there is no API? A successful agent must be capable of navigating an end-user interface. This task is complex, requiring the agent to understand the interface content, whether by processing HTML elements or interpreting pixels in a screenshot. The agent then needs to determine the appropriate action, such as clicking a button or filling out a form, and verify the success of this action by checking for confirmation messages. Each action alters the interface state, influencing subsequent actions and requiring the agent to continuously adapt its approach.

Memory

Memory in AI agents is a critical capability that enables these systems to retain and utilize information from previous interactions or tasks. This function allows AI to maintain context, learn from experiences, and deliver more coherent and personalized responses.

There are different variations. First, there is short-term memory. It temporarily retains and manipulates information relevant to the immediate task. It tracks recent events or data points needed briefly before being discarded or transferred to long-term memory. Implementation often involves maintaining a log of recent actions or the last few conversational turns.

Then there is long-term memory. At a high level, this type provides the agent with the ability to retain and access information over extended periods, storing accumulated knowledge, learned experiences, and established patterns. It shapes the agent's decision-making processes and adaptability.

Long-term memory is often implemented using vector databases. This allows for efficient retrieval of relevant information based on queries related to events, descriptions, and associated metadata. The structure, representation, and retrieval mechanisms of this data significantly impact the effectiveness of memory recall and the overall performance of the AI agent.

Long-term memory includes

- Episodic Memory: This stores specific events or experiences, allowing the agent to recall past occurrences and apply learned lessons to current situations.

- Semantic Memory: This retains general knowledge and facts about the world, enabling the agent to understand objects, concepts, relationships, and procedures. It provides a broad understanding of the domain, allowing reasoning and inference even in unfamiliar scenarios.

- Procedural Memory: This focuses on storing learned skills and procedures, emphasizing how to perform tasks rather than recalling specific events.

Recent research highlights the effectiveness of these memory systems. For instance, a study demonstrated how AI agents with short-term, episodic, and semantic memory systems outperformed those without such structured memory in complex environments. This highlights the benefits of these memory types for task performance and learning efficiency (Kim et al., 2023).[10]

The development of advanced memory systems in AI agents is crucial for their ability to handle increasingly sophisticated tasks with greater independence. For example, the JARVIS-1 agent uses multimodal memory to enhance its task planning and execution in complex, open-world environments, demonstrating significant advantages in performance and adaptability (Weng, 2023).[11]

Planning

Planning in AI agents involves leveraging LLMs to autonomously determine a sequence of steps necessary to achieve a broader objective. This process allows AI to break down complex goals into manageable tasks. This enhances its capability to execute intricate projects. For example, an LLM can guide an AI agent in organizing a virtual event by breaking the task into smaller steps such as selecting speakers, scheduling sessions, and coordinating technical support.

Recent advancements illustrate the profound impact of LLM-based planning on autonomous agents. The Reflexion framework, for instance, combines planning, self-reflection, and memory to iteratively enhance task performance. This allows agents to dynamically adjust their plans based on feedback and previous experiences. This helps to improve decision-making and execution over time.

---

[10] https://ojs.aaai.org/index.php/AAAI/article/view/25075
[11] https://ar5iv.labs.arxiv.org/html/2311.05997

Furthermore, the TPTU (Task Planning and Tool Usage) framework emphasizes the synergy between planning and tool usage. This framework evaluates how effectively LLMs can plan tasks and use tools. AI agents can either adopt a one-step approach, which outlines the entire task at once, or a sequential approach, which addresses each subtask individually, allowing for ongoing feedback and adjustments.

In practical scenarios, planning enables AI agents to manage tasks that require dynamic responses and specialized knowledge. For example, an AI agent tasked with automating a home garden can plan steps such as setting up sensors, configuring irrigation schedules, monitoring plant health, and integrating data with a smartphone app.

While planning significantly enhances AI capabilities, it also introduces unpredictability, as agents might deviate from expected behaviors due to the complexity of generating dynamic plans. However, with ongoing advancements in this field, the reliability and sophistication of planning in AI agents are anticipated to improve.

Multi-agent Collaboration

Multi-agent collaboration uses various LLMs that work together to accomplish complex tasks. This approach is similar to how human teams operate—that is, each agent specializing in different subtasks to achieve a common goal. For example, in a marketing campaign project, different AI agents could assume roles such as content creator, market analyst, campaign strategist, and performance evaluator.

By prompting one or multiple LLMs to perform distinct tasks, you can create specialized agents. For instance, in a marketing campaign, an agent tasked with content creation might be prompted with instructions like, "You are an expert in crafting engaging marketing copy. Write content for the campaign focused on promoting the new product...." This method leverages the strengths of LLMs while maintaining a clear focus on specific subtasks, enhancing overall performance and efficiency.

Another agent could be assigned to market analysis with a prompt such as, "You are skilled in analyzing market trends and consumer behavior. Provide insights based on the latest data to inform the campaign strategy."

Research has shown that multi-agent systems often outperform single-agent setups. Studies like those from MIT demonstrate that collaborative interactions among multiple AI models can significantly improve reasoning and factual accuracy.[12] By engaging in deliberative processes, these agents can critique each other's outputs, leading to more accurate and comprehensive solutions.

Autonomy

AI agents exhibit autonomy by independently making decisions and executing tasks without constant human intervention. This autonomy stems from their ability to process data, learn from experiences, and adapt to new situations in real time. Advanced algorithms and machine learning techniques enable these agents to evaluate their environments, recognize patterns, and predict outcomes, allowing them to take actions that align with their programmed goals. For instance, in autonomous vehicles, AI agents must constantly interpret sensor data to navigate roads, avoid obstacles, and make driving decisions that ensure safety and efficiency. These decisions are made on the fly, showcasing the agents' ability to function autonomously in dynamic environments.

Moreover, AI agents enhance their autonomy through continuous learning and adaptation. Machine learning models allow agents to learn from their experiences and improve their performance over time. This

---

[12] https://news.mit.edu/2023/multi-ai-collaboration-helps-reasoning-factual-accuracy-language-models-0918

learning process involves analyzing past actions and outcomes to refine future strategies. For example, in customer service applications, AI agents can learn from previous interactions to provide more accurate and personalized responses in subsequent engagements.

However, it is often unwise to have a completely autonomous AI agent. Instead, there is a spectrum of autonomy and control that should be considered. Human oversight remains crucial in many scenarios to ensure that AI agents' actions align with broader ethical standards, safety protocols, and organizational goals. By balancing autonomy with human control, we can leverage the strengths of AI while mitigating risks associated with unsupervised decision-making.

Yes, there is much that goes into an agent. But this does not imply that you need to use all the components. You may need only a couple. It depends on the use case.

UI and UX

The user interface (UI) and user experience (UX) are crucial components of software applications. They directly impact user satisfaction, engagement, and productivity.

A well-designed UI ensures that the software is visually appealing and intuitive, making it easier for users to navigate and accomplish their tasks efficiently. Good UX design, on the other hand, focuses on the overall experience users have with the application, including ease of use, accessibility, and responsiveness. Together, UI and UX design help reduce the learning curve for new users, minimize errors, and enhance the overall effectiveness of the software.

This not only boosts user satisfaction but also drives higher adoption rates and customer loyalty. A study by Forrester Research found that a well-designed UI could increase a website's conversion rate by up to 200%, while better UX design could yield conversion rates up to 400%.[13]

As AI agents evolve, rethinking UI and UX design becomes essential to deal with the unique challenges posed by LLMs. Given that LLMs are not always perfect and can sometimes be unreliable, traditional chat interfaces have been an early approach. This interface allows users to easily see the AI's actions, receive streamed responses, correct the AI by responding to it, and ask follow-up questions. This interactive and transparent format ensures that users can remain in control and make necessary changes.

However, there are limitations to this approach. The human remains very much in the loop, making the system more of a copilot rather than an autonomous operator.

One way to address this balance is by ensuring transparency and accountability in the AI's actions. For instance, in a home automation scenario, having a detailed log of everything the agent has done allows users to review and modify actions if necessary.

This review process could be streamlined through an interface that lets users easily modify the schedule for devices like lights, thermostats, and security systems. The AI can autonomously manage these devices, but users can still step in to adjust settings or provide feedback, which the AI can then learn from and adapt to in future tasks.

Moreover, the interface for interacting with AI agents can be designed to be more proactive and integrated into everyday devices. Instead of requiring users to open an application, the AI could work in the background and periodically reach out with updates or queries. For example, an AI agent might notify you through your smart home hub or

---

[13] https://www.forrester.com/report/The-Business-Impact-Of-Customer-Experience-Q4-2016/RES137870

wearable device with a message like, "Your energy consumption is higher than usual today. Would you like me to adjust the thermostat settings to save energy?"

This proactive approach ensures that AI agents are seamlessly integrated into users' lives, providing assistance as needed without requiring constant manual engagement.

Ultimately, rethinking UI and UX for AI agents involves creating systems that are both user-friendly and capable of operating with a degree of autonomy while maintaining transparency and reliability. This ensures that users can trust AI agents to handle tasks efficiently, intervening only when necessary to ensure the desired outcomes.

New Approaches to Development

Traditional software development follows a fairly deterministic workflow. It is based on a structured and sequential approach to creating software applications. This process typically begins with requirement analysis, where the needs and objectives of the software are clearly defined. This is followed by system design, where the architecture and detailed specifications are created. Next comes implementation or coding, where developers write the actual code according to the design specifications. Once the coding is complete, the software undergoes rigorous testing to identify and fix any bugs or issues. After successful testing, the software is deployed into the production environment. Finally, maintenance and updates are performed as necessary to address any issues that arise after deployment.

The deterministic nature of traditional software development lies in its predictability and repeatability. Each phase of the development process is well-defined and follows a linear progression. The clear documentation and structured processes make it easier to manage large teams and complex projects.

Developing generative AI agents significantly differs from traditional software development due to its reliance on probabilistic outcomes rather than deterministic processes. This can be a major adjustment for developers.

Let's take a look at a typical workflow. The first step is to identify the use case, a task that can be complex since certain scenarios may not be suitable for AI due to the need for predictability. Once a suitable use case is determined, selecting one or more models is the next challenge. This selection process is intricate because models are sophisticated and frequently updated.

Cost is another critical factor in developing generative AI agents. Whether using an API or running models locally, the expenses can be substantial. Running a model locally may require buying costly hardware, such as GPUs. Furthermore, the complexity of the workflows must be thoroughly evaluated. Given that LLMs operate on probabilities, there is always the risk of incorrect outputs or decisions. To mitigate these risks, implementing guardrails and considering options for a human-in-the-loop are common practices to ensure safety and accuracy.

Testing generative AI agents presents its own set of challenges due to the unpredictability of the responses. This testing phase can be lengthy and detailed, requiring extensive trials to ensure reliability and effectiveness.

According to Sonya Huang and Pat Grady, who are partners at Sequoia Capital:

Existing monitoring tools don't provide the level of insights you need to trace what went wrong with an LLM call. And testing is different in a stochastic world, too—you're not running a simple "test that 2=2" unit test that a computer can easily verify. Testing becomes a more nuanced concept with

techniques like pairwise comparisons (e.g. Langsmith, Lmsys)
and tracking improvements/regressions. All of this calls for a
new set of developer tools.[14]

To improve accuracy, it is often necessary to use databases with proprietary information, adding another layer of complexity. This may involve fine-tuning the model or employing techniques like Retrieval-Augmented Generation (RAG) to enhance the model's performance. Each of these steps underscores the dynamic and adaptive nature of developing generative AI agents. This certainly highlights the differences from the more deterministic workflows of traditional software development.

Flavors of AI Agents

AI agents come in two primary forms: embodied agents and software agents. Each type serves distinct purposes and operates in different environments. They leverage the unique capabilities of AI to address specific needs and challenges.

Embodied agents are AI systems that interact with the physical world or simulated 3D environments. These agents are often used in robotics, where they can perform tasks such as assembly line work, warehouse management, and autonomous navigation. In video games, embodied agents control non-player characters (NPCs), creating more immersive and realistic experiences for players. The development of embodied agents requires sophisticated algorithms that enable perception, decision-making, and action within dynamic environments. These agents often rely on sensors, cameras, and other input devices to gather information about their surroundings, process this data in real time, and execute appropriate actions.

---

[14] https://www.sequoiacap.com/article/goldilocks-agents/

Software agents, on the other hand, operate within digital environments, handling tasks related to office work, workflows, and data management. These agents can automate repetitive tasks, manage emails, schedule appointments, and facilitate complex business processes. Software agents are designed to improve productivity and streamline operations by acting as intelligent assistants that can understand and execute various commands based on user inputs.

The development of both embodied and software agents involves distinct challenges and methodologies. Embodied agents require extensive training in real or simulated environments to handle physical tasks effectively. This training often involves reinforcement learning, where agents learn through trial and error to optimize their actions. Conversely, software agents are typically trained on large datasets using LLMs to understand and generate humanlike responses.

As for this book, the primary focus will be on software agents.

Brief History

AI agents have been around since the dawn of AI, with early programs in the 1950s laying the groundwork for their development. The Logic Theorist (1955), created by Allen Newell and Herbert A. Simon, was among the first AI programs, designed to mimic human problem-solving skills by proving mathematical theorems from *Principia Mathematica*. Its use of automated reasoning and heuristics showcased the potential for machines to perform intelligent tasks. Following this, Newell and Simon developed the General Problem Solver (1957), a more versatile system capable of applying general strategies to solve a wide range of problems. Introducing means-end analysis and hierarchical problem-solving, GPS aimed for universal applicability, influencing both AI and cognitive psychology. These foundational efforts demonstrated that machines could emulate human reasoning and inspired future AI advancements.

Of course, generative AI agents represent a very recent development in the field of artificial intelligence. The breakthrough came with the launch of OpenAI's ChatGPT in November 2022, which rapidly became the fastest-growing web application.

OpenAI's subsequent models, including GPT-4o, have significantly advanced generative AI's capabilities, enabling more accurate and sophisticated text generation, reasoning, and content creation. These developments have allowed AI to assist in diverse applications, from customer service to software development.

LangChain has played a key role in the development of generative AI agents by providing a framework that simplifies the integration LLMs with various data sources and tools. This technology emerged around mid-2023, when it began offering comprehensive support for agents that can plan, execute tasks, and adapt based on outcomes.

In the meantime, other systems like BabyAGI and AutoGPT emerged to build generative AI agents. They initially generated significant buzz within the AI community. BabyAGI, created by Yohei Nakajima, and AutoGPT, developed by Toran Bruce Richards, promised revolutionary capabilities by leveraging LLMs like OpenAI's GPT-4 to automate complex tasks with minimal human intervention. However, the initial excitement was soon tempered by the realization of their limitations. Both systems struggled with brittleness and generalization, often getting stuck in loops or failing to follow through on tasks coherently.

But this was OK. This is a normal part of the innovation process. There are often false starts, and these initial attempts help identify critical areas for improvement. The experiences with BabyAGI and AutoGPT provided valuable lessons and insights that contributed to the refinement and evolution of autonomous AI agents.

New platforms like LangGraph, AutoGen, and CrewAI are now leading the way in this ongoing evolution. LangGraph provides a framework for building stateful, multi-agent systems that can handle complex workflows and integrate seamlessly with various tools, enhancing the reliability and efficiency of AI agents. AutoGen offers advanced capabilities for generating AI-driven content and automating tasks with greater precision and adaptability, leveraging the latest advancements in machine learning and natural language processing. CrewAI focuses on collaborative AI, enabling multiple agents to work together on intricate projects, optimizing resource utilization, and improving overall performance. These platforms, which are open source, represent the next step in the journey of generative AI, building on past experiences to create more resilient and versatile AI agents.

Emerging proprietary systems are also making significant strides, especially in enterprise-grade applications. These systems are designed to meet the complex needs of businesses, offering robust security, scalability, and integration capabilities. Companies like Microsoft and Google are integrating advanced AI functionalities into their enterprise solutions, providing tools that enhance productivity, automate routine tasks, and deliver actionable insights across various business functions.

Again, this is early days. But the pace of innovation and investment in core technologies for AI agents has remained brisk.

LLMs, Copilots, and RPA

Generative AI agents differ from general-purpose LLMs like ChatGPT, Claude, and Gemini in several key aspects. While LLMs excel in generating text based on prompts and can access tools like Internet searches or APIs for additional information, they typically do not engage in complex actions or planning. These LLMs are primarily designed for conversational interactions and do not possess the specialized capabilities or domain-specific knowledge that generative AI agents often require. As they evolve, LLMs are incorporating more agentic features, but their primary function remains centered around providing information and engaging in dialogue rather than executing tasks or making decisions.

Then what about copilots? They are more specialized and task oriented. These agents are tailored for specific applications or domains such as marketing, law, or HR. For instance, a marketing copilot might assist in drafting ad copy or analyzing campaign performance data. They are capable of not only generating text but also retrieving and integrating relevant information from various sources, such as emails or databases, to enhance their output. Users can interact with these agents to receive suggestions, which they can accept, reject, or modify. This helps to streamline workflows and enhance productivity within specific professional contexts.

Robotic process automation (RPA), on the other hand, represents a different category altogether. RPAs are focused on automating repetitive and rule-based tasks traditionally performed by humans. They operate based on predefined rules and structured data, mimicking human actions like clicking interfaces or entering data into forms. RPAs do not utilize AI for decision-making but can be integrated with AI agents to enhance their capabilities. This integration allows RPAs to handle more complex tasks that require cognitive functions such as natural language understanding or pattern recognition, expanding their utility beyond basic automation.

However, over time, there will likely be a move mainly toward AI agents. So ultimately, there may not be much distinction between LLMs, copilots, and RPA.

Note   The software-as-a-service (SaaS) market is valued at about $261.15 billion.[15] But AI agents are positioned to revolutionize this sector. One major area is with the traditional subscription business model. After all, if an AI agent can handle many of the activities of a certain role and there is likely to be minimal interaction with a

---

[15] https://www.grandviewresearch.com/industry-analysis/
saas-market-report

human worker, then why tie the charge for the software according to the number of seats or users? Rather, it seems more likely that the business model will be based on the measurable improvements to productivity, cost savings, and effectiveness of the decision-making. It is essentially an outcome-based pricing model.

Use Cases

Sandi Besen is an Applied Artificial Intelligence Researcher at IBM. "We focus on staying 6 months ahead of the AI curve by experimenting with emerging AI technology and how we can apply it to enterprise solutions," she said. "As of the last 6 months our focus has been solely on AI Agents."[16]

What she has noticed is that there has been a shift to embedding agents into operational processes of business—rather than as being a copilot. This has meant the emergence of a wide array of use cases.

According to her:

> Some use cases that our clients have been interested in across different industries are: helping agents handle flight the rebooking of passengers when flights are cancelled, having agents find conflicting arguments in new policies that directly contradict existing policies, creating long form documentation with validation and fact checking, research agents that pull information from many sources to help complete a task, which is an exhaustive search that would take humans too long.

So in the rest of the chapter, we'll take a look at how companies are deploying AI agents.

---

[16] From an interview with the authors.

Sierra

Bret Taylor has an impressive career in the technology sector, beginning with his contribution to the creation of Google Maps. Following this achievement, he ventured into entrepreneurship and cofounded FriendFeed, a social media aggregation company later acquired by Facebook. The "like" feature from FriendFeed was integrated into Facebook, where Taylor eventually served as the Chief Technology Officer.

In 2012, Taylor left Facebook to start Quip, a productivity tool designed to compete with Google Docs. Quip was eventually sold to Salesforce.com for a substantial amount, and Taylor rose to the position of co-CEO at Salesforce.com.

Along the way, he saw the growing importance of AI. So to capitalize on this trend, he cofounded Sierra with Clay Bavor, Google's former Head of VR.

Sierra focuses on creating AI agents aimed at enhancing customer experience for enterprise clients. The platform emphasizes high levels of security, governance, and privacy, offering QA and audit tools. Notable clients include Weight Watchers, Sonos, and OluKai. Taylor highlighted the importance of democratizing access to advanced technology, stating, "The greatest opportunity we have is to enable every company, no matter how sophisticated or technical, to deploy AI successfully."[17]

Sierra's AI agents are designed to integrate with existing business infrastructures, leveraging company data to inform actions, which must be formally approved. These agents are sophisticated enough to handle complex customer interactions with empathy. The platform uses a multi-model approach, sometimes employing up to seven models, including a "supervisor" model that monitors and ensures the quality of responses.

---

[17] https://www.cxtoday.com/contact-centre/ex-salesforce-co-ceo-launches-ai-agent-startup/

Furthermore, the founders have innovated in their pricing strategy, opting for an outcome-based model where customers pay only when problems are resolved, rather than traditional subscription or usage-based fees.

Sierra has successfully raised $110 million in funding from Sequoia and Benchmark, showcasing strong investor confidence in their vision and capabilities.[18]

Enso

Mickey Haslavsky's parents ran small businesses, and from a young age, he witnessed the challenges they faced, particularly with adopting new technologies. Inspired by these experiences, Haslavsky founded Enso, a company that leverages AI agents to assist small- and medium-sized businesses (SMBs). He noted:

> *However, I realized the challenge extends beyond generational divides; small businesses, overwhelmed with managing everything on their own, often struggle to embrace digital tools and advancements due to limited bandwidth and resources.*[19]

Enso's AI agents are designed to be user-friendly, requiring no technical skills. They operate mostly in the background, handling various tasks and processes, with users primarily needing to make approvals and minor adjustments. These agents are built on extensive API integrations, leveraging large language models (LLMs) and robotic process automation (RPA). They are trained across multiple industries, including healthcare, financial services, and beauty, to provide versatile support in areas like marketing, content creation, and research.

---

[18] https://www.cxtoday.com/contact-centre/ex-salesforce-co-ceo-launches-ai-agent-startup/
[19] https://enso.bot/blog/the-story-behind-enso

For example, Enso's AI can generate a podcast by using a research tool to find relevant topics, LLMs to create and fine-tune the script, voice-generation tools to record voices, AI music generation for the intro and outro, and video editing tools to finalize the video. Enso's services are priced between $29 and $79 per month.

In July 2024, Enso announced a seed funding round of $6 million, led by NFX and supported by various angel investors, including Yossi Matias, head of AI at Google Research, and Shmil Levy, former general partner at Sequoia Capital.[20]

Asana

Asana is a web and mobile application designed to help teams organize, track, and manage their work. It was founded in 2008 by Dustin Moskovitz and Justin Rosenstein, both former Facebook employees. The company has developed an AI agent system called "AI Teammates," which underscores the importance of having humans in the loop.[21]

AI Teammates allow organizations to create customized agents to manage workflows. This is significant because workflow tools are often rigid and can easily break if processes change. For instance, if a help ticket is submitted with missing or inadequate information, the AI teammate can return it to the submitter, requesting the necessary details. This might involve using generative AI to assist the employee in writing a complete ticket before sending it back to the AI teammate, which can then route it to the appropriate person for resolution.

---

[20] https://enso.bot/blog/press-release-enso-launches-first-guided-ai-agents-platform-with-6m-in-funding-from-nfx
[21] https://techcrunch.com/2024/06/05/asana-introduces-ai-teammates-designed-to-work-alongside-human-employees/

One of Asana's key advantages is its extensive datasets, with over 100,000 customers. The Asana Work Graph tracks intricate connections and relationships, enabling the AI to understand not just how work happens, but how it happens in specific instances. This means that when AI Teammates are embedded into a workflow, they are given specific tasks and know which information to access, increasing their likelihood of performing the right actions.

Note   Aaron Levie, who is the cofounder and CEO of Box, has said: "With AI agents, there are multiple components to a self-reinforcing flywheel that will serve to dramatically improve what AI agents can accomplish in the near and long-term: GPU price/performance, model efficiency, model quality and intelligence, AI frameworks and infrastructure improvements. Altogether, these improvements mean that what we're seeing today in AI agents is barely scratching the surface of what will soon be possible. Something that may not work today or seems too expensive to automate, may just be one performance improvement cycle away from becoming possible."[22]

Conclusion

The evolution of AI agents marks a significant milestone in the AI landscape, characterized by rapid advancements and widespread optimism among industry leaders. Andrew Ng and other visionaries like

---

[22] https://www.linkedin.com/posts/boxaaron_ai-agent-innovation-is-going-to-follow-the-activity-7201264007775756289-Kbt5/?utm_source=share&utm_medium=member_ios

Bill Gates highlight the transformative potential of AI agents, envisioning a future where these systems become integral to both personal and professional spheres.

For software developers, in particular, this represents a significant opportunity. The ongoing advancements promise to redefine traditional workflows and software paradigms, offering developers a chance to innovate and create next-generation applications. This new wave of AI technology will demand fresh approaches and solutions, making it an exciting time for those looking to push the boundaries of what software can achieve.

CHAPTER 2

Generative AI Foundations

Generative AI is a branch of artificial intelligence that provides for the creation of diverse content such as blogs, articles, code, images, videos, and music. This process is initiated by entering a prompt into a chatbot system, which then generates humanlike output based on the given instructions. One of the primary ways to access generative AI is through a large language model (LLM). An LLM is a sophisticated system trained on vast amounts of data across various topics, including biology, marketing, history, finance, medicine, technology, literature, entertainment, and more. This extensive training allows the model to perform tasks such as language translation, classification—where it categorizes data into predefined groups—and summarization.

Understanding generative AI is crucial for developing AI agents, as it provides insight into the capabilities and limitations of this technology. By grasping how generative AI works, one can better harness its potential while being mindful of its constraints. This chapter will look into the fundamentals of generative AI, exploring its applications and significance in AI agent development.

© Tom Taulli, Gaurav Deshmukh 2025
T. Taulli and G. Deshmukh, *Building Generative AI Agents*,
https://doi.org/10.1007/979-8-8688-1134-0_2

Note   Chatbot technology is nothing new. The roots go back to the 1960s. Joseph Weizenbaum, a professor at MIT, created ELIZA. It was essentially a virtual therapist. A user could enter a question—in a teletype interface—and the chatbot would provide a response. True, it was mostly a mimicking of the question. But ELIZA still proved to be quite powerful. Some users actually thought it was a real person.

Pretrained Models

LLMs are pretrained. This means they are trained on vast amounts of data, often drawn from extensive corpora like Wikipedia, Reddit, and many other sources across the Internet. The speculation is that some of massive LLMs encompass most of the content available online, enabling them to generate highly informed and contextually relevant outputs. Additionally, some pretrained models incorporate proprietary information, as companies like OpenAI have licensed content from publishers such as News Corp, Springer, and Vox.

During the pretraining phase, the models learn embeddings, which are dense vector representations of words or tokens. These embeddings, essentially fixed-length arrays of numbers, capture semantic relationships between words, allowing the model to understand and generate humanlike text based on context.

Traditionally, AI models were trained on structured, labeled data, which made it easier to process and categorize information. However, generative AI distinguishes itself by allowing training on unstructured data. This flexibility enables the models to handle a more diverse range of inputs and generate outputs that closely mimic human language. The effectiveness of these models is further enhanced by scaling laws, which posit that the performance of an LLM improves with the increase in the quantity of training data and the size of the model itself.

An important development in the field of pretrained models is the emergence of domain-specific LLMs. These models are fine-tuned with data from specific industries or fields, such as legal documents, medical records, or financial reports, making them highly specialized for particular tasks. This specialized training allows them to better understand and process the unique language and concepts of their domain, enhancing their accuracy and relevance. As a result, domain-specific LLMs provide more precise and tailored outputs compared to general-purpose LLMs, which may struggle with domain-specific terminology and nuances.

Despite their advantages, pretrained models come with certain drawbacks. One significant limitation is the cut-off date for the training data, meaning that an LLM will lack the most recent information. Additionally, there are concerns about the availability and quality of useful data, with some believing that companies are exhausting the reservoir of valuable data. As more Internet content becomes AI generated, there is a risk of creating a corrosive feedback loop that could degrade the quality of the models.

To mitigate these challenges, an emerging trend involves generating synthetic data. Synthetic data refers to artificially created data that mimics real-world data but is generated through algorithms rather than collected from actual events. This approach reduces the dependence on public Internet content and proprietary data, potentially lowering costs and circumventing licensing issues. Synthetic data generation could thus play a crucial role in sustaining and enhancing the performance of future LLMs.

Transformer Models

The transformer model represents a revolutionary shift in the field of natural language processing (NLP) due to its unique architecture and innovative use of attention mechanisms. Introduced in the seminal 2017 paper "Attention Is All You Need," the transformer model has significantly outperformed previous models—like recurrent neural networks (RNNs)—by efficiently processing large datasets. Unlike traditional models that rely on sequential data processing, the transformer employs parallel processing, allowing it to handle vast amounts of text data more quickly and accurately. This also makes these types of models ideal for GPUs.

A key component of the transformer model is its attention mechanism, specifically the self-attention mechanism. This allows the model to weigh the importance of different tokens within a given context. This helps to understand the relationships and dependencies between them. By doing so, it can capture the nuanced meaning of words based on their surrounding context. For example, the word "bark" can refer to the sound a dog makes or the outer layer of a tree, and the transformer model can discern the correct meaning based on the other words in the sentence.

There are three primary types of transformer models, each with distinct applications. Autoregressive language models, like OpenAI's GPT series, predict the next token in a sequence based solely on the preceding tokens. These models are excellent for tasks involving text generation.

Autoencoding language models, such as Google's BERT, take a different approach by predicting tokens based on the surrounding context, making them bidirectional. This bidirectional nature allows these models to excel in tasks like text classification, sentiment analysis, and named entity recognition. They effectively understand the full context of a sentence. This helps to improve the accuracy in understanding the meaning and intent behind the text.

The third type of transformer model combines both autoregressive and autoencoding techniques. An example of this is the T5 model, which can be fine-tuned for various tasks, leveraging the strengths of both approaches to achieve state-of-the-art performance across a range of NLP applications.

In addition to these foundational types, transformer models have evolved to incorporate various enhancements, such as sparse attention mechanisms, which reduce computational complexity.

Transfer Learning

Transfer learning is a machine learning technique where a model trained on one task is repurposed to perform a different but related task. This method leverages the knowledge gained during the initial training phase to improve performance on the new task.

Transformer models are particularly well suited for transfer learning due to their ability to learn and encode complex language patterns during pretraining. These models are initially trained on extensive and diverse text corpora using unsupervised learning methods, learning to understand syntax, semantics, and the contextual relationships between words. Once pretrained, these transformer models can be fine-tuned for specific downstream tasks, such as text classification, sentiment analysis, or question answering. Fine-tuning involves adjusting the model's parameters using a smaller, task-specific dataset, allowing the model to adapt its general language understanding to the nuances of the particular task, leading to enhanced performance.

Note   Parameters are variables that the model learns and adjusts during the training process to make accurate predictions. In the context for transformer models, these parameters include weights and biases that help the model understand the complex patterns and relationships in the data. For example, in an LLM, parameters determine how strongly each word in a sentence is connected to other words. This helps the model capture context and meaning.

Retrieval-Augmented Generation (RAG) extends this approach by combining the capabilities of pretrained language models with an external knowledge retrieval system. RAG enhances performance on specific tasks by retrieving relevant information from a large corpus and integrating it into the generation process. This makes RAG particularly useful for tasks requiring current knowledge or detailed context, further improving the adaptability and efficiency of transformer models in various NLP applications.

The transfer learning approach with transformers offers several advantages. It is efficient, as the model has already learned a vast amount of information during pretraining, requiring significantly less data and time to fine-tune for specific tasks. Pretrained transformers often achieve state-of-the-art results on various NLP tasks after fine-tuning. Additionally, leveraging pretrained models reduces the computational resources required, making it feasible to train high-performing models even with limited data and hardware.

Alignment in Language Models

Alignment in language models refers to the model's ability to produce responses that meet the expectations and requirements of the user. This involves ensuring that the model's outputs are coherent, contextually appropriate, and align with the desired goals of the user.

Reinforcement Learning from Human Feedback (RLHF) is a popular method used to enhance the alignment of LLMs. RLHF involves using feedback from human evaluators to fine-tune the model's performance. Instead of relying solely on traditional supervised learning, which can be limited by the quality and scope of labeled data, RLHF allows the model to learn from a smaller, high-quality batch of human feedback on its outputs. This iterative process helps the model better understand and meet user expectations, leading to more accurate and satisfactory responses.

The implementation of RLHF has shown significant improvements in modern LLMs. By incorporating human feedback, these models can refine their responses to be more relevant and contextually appropriate. This method helps address issues like generating safe and ethical content, reducing biases, and improving overall user satisfaction.

While RLHF has proven effective, other approaches are also being explored to achieve better alignment in LLMs. One such method is Reinforcement Learning with AI Feedback, demonstrated by constitutional AI. This approach uses feedback from AI systems to guide the training and fine-tuning of language models. By leveraging both human and AI feedback, researchers aim to create models that are even more aligned with user expectations and capable of producing high-quality outputs consistently.

Multimodal LLMs

Multimodal LLMs are a cutting-edge development in artificial intelligence. They are designed to process and generate information across various data types such as text, images, audio, and video. These models transcend the limitations of traditional, text-only language models by integrating different forms of input and output, enabling more nuanced and versatile interactions. This advancement aligns with the broader objective of AI to mirror humanlike understanding and interaction, which naturally involves multiple senses.

Recent strides in multimodal LLMs have marked significant milestones in AI research. For example, OpenAI's advancements with GPT-4o and Sora highlight the importance of incorporating additional modalities, such as images, into language models, viewing this integration as a critical frontier in AI development. Other notable models include DeepMind's Flamingo and Microsoft's KOSMOS-1, which are leading the charge in this multimodal shift.

One of the primary advantages of multimodal LLMs is their ability to process and interpret different types of data simultaneously. This capability leads to more informed decision-making and richer user experiences. For instance, in ecommerce, these models can enhance product recommendations by understanding both visual product attributes and textual customer reviews.

However, deploying multimodal LLMs comes with challenges, including the complexity of training these models and ethical considerations. Training requires extensive and diverse datasets, along with significant computational resources. Furthermore, the integration of different data types must be managed carefully to ensure coherent and useful outputs.

Types of Models

The first OpenAI GPT model, launched in 2018, had 117 million parameters and marked a significant breakthrough in natural language processing. Known as GPT-1, it utilized a transformer architecture and was trained on a large corpus of Internet text. Although limited in its capabilities, GPT-1 demonstrated the ability to generate coherent and contextually relevant text.

Following the release of GPT-1, OpenAI rapidly advanced its language model technology with subsequent iterations. In 2019, GPT-2 was introduced with 1.5 billion parameters, significantly improving performance. The leap to GPT-3 in 2020 was even more dramatic, featuring 175 billion parameters. This exponential increase in model size enabled GPT-3 to perform a wider range of tasks with remarkable accuracy.

However, developing these advanced models required enormous computational resources and financial investment. Training such large-scale models necessitated powerful hardware and extensive data. Recognizing the potential and resource demands of these models, Microsoft made a $1 billion investment in OpenAI in 2019. This partnership enabled the scaling and deployment of these advanced language models, driving forward the progress of AI and its applications across various industries. Over the years, Microsoft's investment would total more than $13 billion.

With the introduction of GPT-4, OpenAI has not disclosed the exact number of parameters, but rumors suggest that it exceeds 1 trillion parameters. This lack of transparency has become common among other developers of LLMs as they seek to protect their innovations in a highly competitive field.

Today, the landscape for LLMs is diverse, with new models being launched or upgraded regularly. This rapid pace of development can make it challenging for developers to stay current with the latest advancements.

There are different types of LLMs. They include proprietary systems and open source projects. There are also much smaller platforms, which are called small language models (SLMs).

Let's take a look at each of these.

Proprietary LLMs

Proprietary LLMs are advanced AI systems developed, owned, and controlled by private organizations. Examples include OpenAI's GPT models, Anthropic's Claude, and Google's Gemini.

Here are some of the advantages of proprietary LLMs:

- Financial Resources and Top Talent: Significant financial investments allow these companies to attract highly experienced data scientists and AI experts, ensuring continuous development and refinement of cutting-edge models.

- Leading Capabilities and Performance: These models are often at the forefront of AI technology. They can offer superior performance for various natural language processing tasks.

- Robust Developer Ecosystem: The availability of well-structured API ecosystems simplifies integration and accelerates the development process. Developers can use common programming languages like Python and benefit from additional tools such as playgrounds for testing and refining applications.

- Cost Efficiency: Using APIs eliminates the need for costly hardware investments.

- Comprehensive Support: The financial strength of these companies enables them to provide extensive customer service and support. This includes access to detailed documentation, tutorial videos, and responsive support teams, ensuring developers can effectively utilize the models and address any challenges they encounter.

- Ethical AI and Responsible Usage: Companies like Anthropic prioritize aligning AI development with human values and safety considerations, addressing concerns related to AI misuse, privacy, and fairness.

- Scalability: The robust infrastructure of these companies supports large-scale deployments and high-demand scenarios, making these models suitable for enterprise-wide implementations and services with high user engagement.

Then what about the downsides? Let's take a look.

While proprietary large language models (LLMs) like OpenAI's GPT models, Anthropic's Claude, and Google's Gemini offer numerous benefits, there are also some notable downsides to consider:

- Limited Customization: Users may have limited ability to modify the models to suit specific needs or to integrate them deeply with unique systems or workflows.

- Data Privacy and Security Concerns: When using proprietary LLMs, sensitive data must be shared with third-party providers, raising concerns about data privacy and security. There is always a risk that data could be mishandled, exposed, or used in ways not intended by the user.

- Dependence on Provider: Relying on a single provider for AI capabilities creates a dependency that can be problematic if the provider experiences outages, changes pricing structures, or discontinues services. This dependency can lead to significant disruptions in business operations. For example, in late 2023, the board of OpenAI abruptly fired the CEO and cofounder, Sam Altman. While he was installed quickly, the event caused much worry among customers. The result was that some looked for alternative models.

- Transparency and Control: Users may not have visibility into how the models are trained, what data is used, or how decisions are made, leading to a lack of control and potential trust issues.

- Competitive Disadvantages: Companies that rely heavily on proprietary LLMs might find themselves at a competitive disadvantage if their competitors have access to similar or better models.

Open Source LLMs and SLMs

Arthur Mensch completed his thesis in 2018 at École Polytechnique and Télécom Paris, focusing on machine learning for functional brain imaging. Following this, he became a researcher at Google's DeepMind, where he concentrated on large language models (LLMs). During his time there, he coauthored a paper on a model called Chinchilla, which challenged the prevailing scaling laws that LLMs had to be massive.[1]

Concerned about the lack of transparency in megatech companies' models, Mensch was inspired to start his own company. In April 2023, he cofounded Mistral with Timothée Lacroix and Guillaume Lample, both of whom were from Meta's AI lab. The company, based in Paris, focused on building more efficient, cost-effective, and open source models.

Mistral's strategy quickly gained traction. Within nine months of its founding, the company attracted some of the world's top investors, raising $500 million at a valuation of $2 billion. This funding would be followed up in June 2023 with a $646 million investment, with a valuation of $6 billion. The investors included Nvidia and Salesforce. At the time, Mistral had only 60 employees.[2]

Mistral's rapid success ignited a frenzy of interest in open source models and small language models (SLMs). Venture capitalists opened up their wallets to fund new deals. In the meantime, developers and researchers flocked to platforms like Hugging Face, which now hosts over 400,000 models. Each day, new models are added, reflecting the relentless pace of innovation.

Open source LLMs offer numerous benefits that are transforming the AI landscape:

---

[1] https://www.wsj.com/tech/ai/french-startup-mistral-ai-raises-650-million-in-bid-to-scale-up-22899937

[2] https://www.wsj.com/tech/ai/the-9-month-old-ai-startup-challenging-silicon-valleys-giants-ee2e4c48

- Transparency: Open source models allow researchers, developers, and users to inspect the architecture, weights, and training data, promoting accountability and enabling the identification and mitigation of biases inherent in the model.

- Community Collaboration: A diverse group of contributors can work on improving the models, fixing bugs, and adding new features. This collective effort accelerates innovation and ensures that the models evolve rapidly to meet various needs.

- Democratization: Open source provides tools that are freely available for use and customization. This accessibility encourages experimentation and the development of new applications across different domains.

- Enhanced Security: With control over the source code, models can be used locally or in private data centers, providing more control over proprietary data and mitigating risks associated with cloud-based deployment.

As for SLMs, they have their own advantages:

- Efficiency: They are designed to deliver robust performance while requiring less computational power, making them suitable for deployment in environments with limited resources. For instance, companies like Apple use SLMs on their smartphones, reducing latency by eliminating the need to access the cloud.

- Tailor the Models: This can be easier with SLMs. There is also usually more flexibility.

- More Focused: They can be optimized for specific
  tasks like summarization or classification without the
  overhead of unused capabilities found in larger models.
  This targeted approach ensures that enterprises
  do not have to maintain models with unnecessary
  functionalities. After all, does an enterprise need the
  capability to do things like write poetry? Of course not.

While open source LLMs and SLMs offer numerous benefits, they also come with several drawbacks that users need to be aware of. In fact, there are still some potential security issues. Downloading them from third-party sites can introduce vulnerabilities, such as model poisoning, where malicious actors embed harmful elements into the models. This risk underscores the need for robust security measures when handling these models.

The cost of using open source LLMs and SLMs is another important consideration. Although the source code is free, running these models requires sophisticated and often expensive hardware. High-performance GPUs or specialized AI accelerators are necessary to efficiently train and deploy these models, which can be cost-prohibitive for smaller organizations. Additionally, the complexity of these models means that skilled professionals, such as data analysts and data scientists, are needed to manage and optimize their use, further increasing operational costs.

Transparency, a touted advantage of open source models, is not always as clear-cut either. While the architecture and weights may be disclosed, the training data often remains undisclosed. This partial transparency can limit the ability to fully understand the model's behavior and potential biases, which is crucial for responsible AI deployment.

Prompt Engineering

When developing AI agents, understanding prompt engineering is essential. Prompt engineering is the key to effectively working with LLMs. It involves crafting inputs that guide these sophisticated AI systems to generate accurate and relevant responses. Mastering this skill ensures that developers can leverage the full potential of LLMs, making interactions with AI more productive and precise.

Prompt engineering is a subfield of machine learning and natural language processing, focusing on enabling computers to comprehend and interpret human language. This process goes beyond merely asking single questions. It often involves an iterative conversation with the LLM, where you refine and adjust your prompts based on the responses until you extract the precise information or answer you need. This back-and-forth interaction helps in honing the model's output to achieve that valuable insight or solution.

There is a lot of hype surrounding prompt engineering, with numerous videos, blogs, and articles claiming to reveal its "secrets." However, it's important to be cautious of these claims. In reality, prompt engineering boils down to a handful of core concepts. Understanding these foundational principles is more valuable than chasing after supposed hidden tricks.

Prompt engineering is a blend of art and science. The fact is, the same prompt can elicit different responses due to the complex probabilistic nature of LLMs. This inherent variability means that trial and error is a common part of the process. Developers must often tweak their prompts multiple times to get the desired outcome, which requires patience and a willingness to experiment.

Moreover, LLMs are frequently updated, which can introduce changes in their capabilities and outputs. These updates can sometimes improve certain aspects while worsening others, adding another layer of complexity to prompt engineering.

41

With all this in mind, let's take a look at some of the key factors for successful prompt engineering.

Be Clear

When working with LLMs, clarity in your prompts is essential for generating accurate and relevant responses. An LLM needs to understand the nuances of your request to offer useful information. By being clear and detailed in your instructions, you set the stage for the model to produce responses that align closely with your expectations.

There are several techniques you can employ to achieve this clarity and context in your prompts.

Details

To obtain a relevant response from an LLM, it's important to provide important details and context in your requests. Failing to do so leaves the model to make assumptions about your intents. A more effective approach is to be specific and include key information in your prompt. For instance, rather than simply asking "Explain gravity," a more effective prompt would be "Explain how gravity affects the orbit of the Earth around the Sun, using simple terms suitable for a high school student." This level of detail guides the model to generate a response tailored to your specific needs.

Persona

A system message is a powerful tool when working with LLMs, as it allows the model to adopt a specific persona or role for its responses. This approach enhances the relevance and quality of the model's output.

System prompt: "When I ask for advice, respond as if you are a seasoned business consultant with over 20 years of experience in the tech industry."

Use Delimiters

Delimiters are valuable tools when working LLMs as they help to clearly demarcate different sections of text that require specific treatment. This technique can enhance the model's focus and accuracy in processing information. Common delimiters include triple quotation marks or section titles.

For instance, when dealing with a document that needs summarization, you could use a prompt like "Summarize the text delimited by triple quotes." You would then enclose the entire document within triple quotes. This approach effectively instructs the LLM to concentrate solely on the content within the specified delimiters, ensuring that it summarizes only the relevant text while disregarding any extraneous information.

Steps for a Task

When working with large language models, breaking down complex tasks into a sequence of explicit steps can improve the model's ability to follow instructions accurately. This approach provides a clear road map for the model to process information and generate responses. Here's an example:

Follow these steps to create a simple Python function:

Step 1: The user will provide a brief description of a function they need. Based on this description, create a function signature, including an appropriate name and parameters. Prefix this with "Function Signature: ".

Step 2: Implement the function body with the necessary logic to accomplish the described task. Use clear variable names and add comments to explain any complex parts. Present the complete function, including the signature from Step 1, with a prefix "Implementation: ".

Step 3: Provide a brief example of how to call the function with sample inputs, and show the expected output. Present this as a code snippet with the prefix "Usage Example: ".

Another approach is to use recursive summarization. This is when you have documents that are too long for an LLM's context window. That is, you can summarize different sections of the document. Then you will summarize the summaries.

Time to Think

When faced with a complex problem or calculation, taking the time to think it through step by step often leads to better results. This principle applies not only to humans but also to LLMs. Models tend to produce more accurate responses when they're prompted to explain their reasoning before providing a final answer. This approach, often called a "chain of thought," allows the model to work through the problem logically, much like a human would.

Incorporating step-by-step reasoning in prompts can enhance problem-solving and decision-making processes. By explicitly asking to "reason it out" or "think things through step by step," we create a mental framework that provides for a more methodical approach. This approach not only improves the quality of responses but also makes the problem-solving process more transparent and easier to follow.

However, other types of prompts can be equally valuable in refining responses. For instance, asking "What can be done to improve this response?" encourages critical evaluation and identifies areas for enhancement. Similarly, prompting with "Are there things I should add?" can reveal overlooked aspects or additional considerations. These prompts, along with others like "What assumptions am I making?" or "What are potential counterarguments?", foster a more comprehensive and nuanced approach to problem-solving. By using a variety of thoughtful prompts, we can guide both human thinking and AI responses toward more thorough, balanced, and insightful outcomes across a wide range of tasks and decisions.

Length of Output

When working with large language models, you have the ability to request outputs of specific lengths. You can specify the desired length in various units such as words, sentences, paragraphs, or bullet points. However, it's important to note that while the model can generally adhere to these requests, its precision varies depending on the unit of measurement. Specifically, asking for a certain number of words may not yield highly accurate results. On the other hand, the model tends to be more reliable when asked to generate a specific number of paragraphs or bullet points.

Summarize the text delimited by triple quotes as one paragraph.
Text

Going Beyond the Transformer

While transformers have significantly advanced generative AI, there is ongoing research to develop more efficient models. Two promising alternatives are Test-Time Training (TTT) and State Space Models (SSMs).

Test-Time Training (TTT) is being developed by researchers from Stanford, UC San Diego, UC Berkeley, and Meta. TTT models are designed to process more data than transformers while consuming less energy. Unlike transformers, which rely on a growing lookup table, TTT models use a machine learning model to encode data into representative variables called weights. This method ensures that the model size remains constant, no matter how much data it processes. TTT can be applied to various data types, including text, images, audio, and video.

Another alternative, State Space Models (SSMs), are also being explored for their computational efficiency and scalability. Like TTT, SSMs can handle larger datasets more efficiently than transformers.

Despite their potential, both TTT and SSM technologies are still in the early stages of development. For now, transformers remain the default model for generative AI, but ongoing research aims to improve these systems' efficiency and capabilities, potentially leading to new models that could eventually surpass transformers.

Conclusion

As we conclude this chapter on the foundations of generative AI, it's clear that this technology represents a monumental leap in artificial intelligence. By enabling the creation of diverse content, from text and images to music and video, generative AI has unlocked new possibilities across industries. The backbone of this technology, LLMs, exemplifies the advancements in AI, showcasing the ability to generate humanlike outputs based on extensive training data. This understanding forms the bedrock upon which AI agents are developed, allowing them to perform tasks that were once considered the exclusive domain of humans.

Reflecting on the evolution of generative AI, we see a continuum from early chatbot experiments like ELIZA to today's sophisticated models capable of nuanced and contextually rich interactions. This progress underscores the importance of continuous innovation and adaptation in the AI field. As we look ahead, the potential of domain-specific models and emerging techniques like synthetic data generation and transformer alternatives promises to further enhance the capabilities and efficiency of AI systems.

CHAPTER 3

Types of Agents

Artificial intelligence (AI) agents represent a diverse and evolving field within AI technology. These agents range from simple systems with limited capabilities to sophisticated entities capable of complex decision-making and learning from their environments. The types of AI agents include simple reflex agents, model-based reflex agents, goal-based agents, utility-based agents, and learning agents, each with unique characteristics and applications that cater to different needs and challenges.

However, distinguishing between these different types of agents can be challenging as the technology continues to evolve rapidly. The lines between agent types often blur, making it difficult to categorize them neatly. For example, an agent designed to optimize supply chain logistics might use a combination of goal-based planning and utility-based decision-making to adapt to real-time changes in inventory and demand. This hybrid nature of modern AI agents reflects the complexity and sophistication of current AI applications.

Moreover, as AI technology advances, new capabilities and integration methods emerge, leading to the development of agents that incorporate multiple AI techniques. These hybrid agents can leverage the strengths of various approaches, such as combining the adaptability of learning agents with the strategic planning capabilities of goal-based agents. This convergence makes it increasingly important to understand the foundational concepts of each type to appreciate their specific contributions and potential synergies.

© Tom Taulli, Gaurav Deshmukh 2025
T. Taulli and G. Deshmukh, *Building Generative AI Agents*,
https://doi.org/10.1007/979-8-8688-1134-0_3

In this chapter, we will take a closer look at the main types of AI agents, exploring their unique characteristics, applications, and the nuances that differentiate them.

Simple Reflex Agents

Simple reflex agents are the most basic type of intelligent agents in artificial intelligence. They operate based on a predefined set of rules that dictate how they should respond to specific sensory inputs from their environment. These agents do not store past experiences or use any form of internal memory. Instead, they react solely based on the current percept. Their decision-making process is straightforward, relying on simple "if-then" conditions to determine the appropriate actions.

For instance, in a temperature control system, a simple reflex agent might be programmed with a rule like "If the room temperature exceeds 45 degrees Celsius, then turn on the air conditioning." This rule allows the agent to monitor the temperature through sensors and activate the air conditioning using actuators when the specified condition is met. Such simplicity makes these agents suitable for tasks that do not require complex decision-making or learning from past experiences, such as resetting passwords based on keywords or operating basic physical robots like vacuum cleaners or thermostats that respond to changes in temperature.

One of the main advantages of simple reflex agents is their ease of design and implementation. They require minimal computational resources and can be highly reliable, provided that the sensors are accurate and the rules are well designed. However, they also have significant limitations. These agents are prone to errors if the input sensors are faulty or the predefined rules are not comprehensive enough to handle all possible situations. Furthermore, they struggle with environments that are partially observable or subject to changes that the agents have not been explicitly programmed to handle. This makes them less effective in dynamic or unpredictable environments where more sophisticated decision-making capabilities are necessary.

Model-Based Reflex Agents

Model-based reflex agents are an advanced form of intelligent agents that enhance decision-making capabilities by incorporating internal models of the environment. Unlike simple reflex agents, which react purely based on current percepts, model-based reflex agents consider both the current sensory input and an internal state that represents aspects of the environment that are not immediately observable. In this context, a "percept" is the data or information that an agent receives from its environment via its sensors at a given moment. It represents the current state of the environment as detected by the agent, serving as the basis for the agent's decision-making process. The percept allows the agent to understand and interact with its surroundings by providing real-time input that informs its actions.

The decision-making process of model-based reflex agents involves several steps. First, they perceive the environment through sensors, gathering current state information. Next, they use an internal model to update their understanding of the environment. This internal model includes knowledge of how the environment evolves independently and how the agent's actions can affect it. This model enables the agent to predict probable outcomes before making decisions, allowing for more sophisticated and informed actions.

A key component of these agents is their reasoning mechanism, which evaluates sensor data and the internal model to make decisions. This reasoning can be based on rule-based systems, logical reasoning, or more advanced methods like machine learning and large language models (LLMs). After making a decision, the agent uses actuators to

49

execute actions, which can involve physical movements in robots or virtual operations in software systems. Actuators are components that carry out the agent's decisions by converting them into physical or digital operations that impact the environment. These actions then impact the environment, completing the perception-action cycle.

Model-based reflex agents offer several advantages. They can make quick and efficient decisions by leveraging their internal models to understand the world better. This makes them more adaptable to changes in the environment, as they continuously update their models based on new percepts. However, these agents are computationally expensive, requiring significant resources to maintain and update their models. Additionally, accurately capturing the complexity of real-world environments can be challenging.

A practical use case for model-based reflex agents is in manufacturing systems, where they optimize production processes by predicting machine failures or material shortages. By maintaining a detailed internal model of the production environment, these agents can proactively address issues, improving efficiency and reducing downtime.

Goal-Based Agents

Goal-based agents, also known as rule-based agents, are AI systems designed to achieve specific objectives or goals by considering future outcomes and planning. They share similarities with model-based agents, but goal-based agents have a different approach to decision-making. While model-based agents use historical and current data to make predictions, goal-based agents are driven by specific objectives and use search algorithms to determine the most efficient path to achieve their goals. This involves considering long sequences of possible actions, often referred to as searching and planning, to navigate toward the desired outcome.

The capabilities of goal-based agents extend beyond simple reactionary behavior to proactive planning and optimization. They are future-oriented, using decision-making algorithms to evaluate potential scenarios and adapt their strategies based on new information and changing conditions. This adaptability is essential in environments where the parameters can shift rapidly, such as robotics, autonomous vehicles, and complex game AI.

Goal-based agents are particularly valuable due to their enhanced autonomy, predictive capabilities, and efficiency. They enable systems to function with minimal human intervention by adjusting their actions to meet their goals, predict future scenarios, and find optimal paths to achieve desired results. This not only saves resources and time but also ensures that the system can adapt to new challenges and opportunities as they arise.

Moreover, goal-based agents are utilized in various advanced applications. In generative AI, they are employed for content creation, game design, automated design and prototyping, personalized marketing, intelligent assistants, and financial trading. These agents excel in tasks requiring complex decision-making and strategic planning, making them indispensable in fields that demand high levels of precision and adaptability.

Utility-Based Agents

Utility-based agents are advanced AI systems that use complex reasoning algorithms to achieve the best possible outcomes by evaluating different scenarios. This evaluation process involves a utility function, which assigns values to various states based on their desirability. The agent then selects actions that lead to high-utility states, effectively balancing multiple goals or optimizing specific criteria such as cost, quality, or time.

One of the key strengths of utility-based agents is their ability to adapt to dynamic environments. They continuously reassess their strategies based on new data and changing conditions

This makes them effective many fields. For instance, in financial trading, they help maximize returns by evaluating different investment strategies and their potential outcomes. In logistics, they optimize supply chain operations by balancing cost, delivery time, and resource availability. Additionally, utility-based agents are instrumental in customer service applications, where they can recommend products or services that best meet user preferences, considering factors such as price, quality, and delivery speed.

However, the implementation of utility-based agents comes with certain challenges. These agents require an accurate model of the environment to make reliable decisions. If the model is inaccurate or incomplete, it can lead to suboptimal decision-making and potential errors. Moreover, the computational demands of evaluating multiple scenarios and calculating expected utilities can be resource-intensive, making these agents expensive to operate.

Learning Agents

Learning agents are a cornerstone of artificial intelligence, designed to improve their performance over time by learning from previous experiences. They operate based on sensory inputs and feedback mechanisms, which allow them to refine their actions and decisions dynamically. Initially, learning agents start with basic knowledge, but they adapt and enhance their performance through machine learning techniques. The architecture of a learning agent typically includes four key components: a learning element, a critic, a performance element, and a problem generator. The learning element updates the agent's knowledge base, the critic provides feedback by evaluating the agent's performance, the performance element is responsible for decision-making, and the problem generator introduces new challenges to stimulate continuous learning.

Despite their potential, learning agents have several notable drawbacks. The development and maintenance of learning agents can be costly and resource-intensive. These agents also require large amounts of data to function effectively, which can be a significant limitation in scenarios where data is scarce or expensive to obtain.

Learning agents have diverse applications across various industries, demonstrating their versatility and impact. One prominent use case is personalized recommender systems. By evaluating user behavior and preferences, learning agents power recommendation engines in social networking and ecommerce platforms.

In the healthcare sector, learning agents support medical practitioners by assisting in drug development, individualized treatment planning, medical diagnostics, and patient health data monitoring. By analyzing vast amounts of medical data, these agents help identify patterns and insights that lead to more accurate diagnoses and effective treatments, ultimately improving patient outcomes.

Hierarchical Agents

Hierarchical agents in AI are structured collections of agents organized in a tiered system. In such a hierarchy, high-level agents are responsible for setting overarching goals and constraints, which are then communicated to lower-level agents. These lower-level agents focus on handling specific tasks to achieve the goals set by the higher-level agents. This structure can vary in complexity, with simple systems featuring just two levels—high-level and low-level agents—while more complex systems might include intermediate-level agents to coordinate and manage activities across different tiers.

One of the primary advantages of hierarchical agents is their ability to reduce duplicated efforts and improve resource efficiency. By delegating tasks through a structured hierarchy, decision-making can be faster, as lower-level agents can operate independently within the constraints set by higher-level agents. This streamlined approach allows for more efficient use of computational resources and can lead to quicker responses in executing tasks.

However, there are also downsides to hierarchical agent systems. Fixed hierarchies can limit adaptability, making these systems less effective in dynamic environments where conditions frequently change. This rigidity can prevent the system from adjusting quickly to new circumstances, potentially reducing its overall effectiveness. What's more, hierarchical systems can be challenging to repurpose for different use cases, as their design is often tailored to specific goals and tasks.

Training hierarchical agent systems can also be complex. The need for extensive data labeling and the intricacies involved in defining the interactions and dependencies between different levels of agents can make the training process labor-intensive and time-consuming. Ensuring that each agent, at every level of the hierarchy, functions correctly and aligns with the overall system's goals requires careful planning and substantial computational resources.

A use case for hierarchical agents is in transportation systems, such as managing traffic and routing for logistics. In this scenario, high-level agents could oversee overall traffic flow and set priorities, while lower-level agents handle specific routing decisions for individual vehicles. Intermediate-level agents might coordinate activities within particular regions or districts, ensuring that traffic moves smoothly and efficiently across the entire network. This hierarchical approach can optimize route planning and traffic management, leading to reduced congestion and improved logistics efficiency.

Conclusion

The diverse types of AI agents discussed in this chapter highlight the breadth and depth of capabilities that artificial intelligence can offer. From the simplicity of reflex agents that react to immediate stimuli to the complexity of learning agents that adapt and evolve based on experience, each type of agent serves unique purposes and applications. Understanding these different types is crucial for anyone looking to leverage AI effectively.

As AI technology continues to advance, the distinctions between different agent types may blur even further, giving rise to hybrid agents that combine the strengths of various approaches. This evolution underscores the importance of a foundational understanding of each agent type, as it provides the necessary context to appreciate the advancements and synergies that hybrid agents bring.

CHAPTER 4

OpenAI GPTs and the Assistants API

Sam Altman, who is the CEO and cofounder of OpenAI, has noted that AI will become more ubiquitous in our lives than smartphones. He also said that AI agents would be the next super killer app. According to him, an AI agent will do "[w]hat you really want."[1]

To this end, OpenAI has launched several agentic systems. There are GPTs as well as the Assistants API.

In this chapter, we'll take a look at these tools. But first, we'll describe the OpenAI account and how to get the API key.

Registering for the OpenAI API Key

To use the OpenAI Assistants API, you will need to set up an account with OpenAI. Go to the following URL:

`https://platform.openai.com/docs/overview`

You will select the "Sign Up" button on the top right. You can enter your email address and password. Or, you can create the account using an existing account with Google, Microsoft, or Apple.

You will get a free credit for $5, which is available for three months. After this, you will need to set up your credit card with OpenAI.

---

[1] `https://www.technologyreview.com/2024/05/01/1091979/sam-altman-says-helpful-agents-are-poised-to-become-ais-killer-function/`

© Tom Taulli, Gaurav Deshmukh 2025
T. Taulli and G. Deshmukh, *Building Generative AI Agents*,
https://doi.org/10.1007/979-8-8688-1134-0_4

GPTs

A GPT is a customization of ChatGPT. It involves combining instructions, files, and APIs.

Think of GPTs like the iOS App Store or Google Play. You can find them at the following URL:

```
https://chatgpt.com/gpts
```

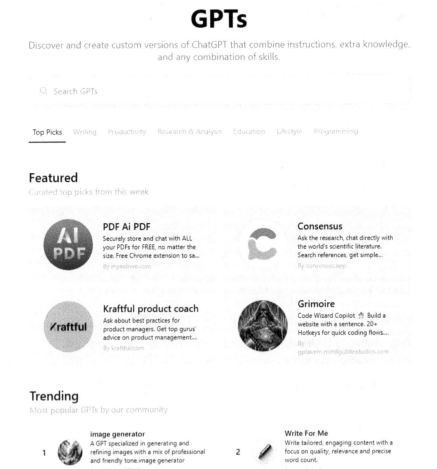

Figure 4-1. This is the main screen for OpenAI's GPTs, which are customizations for ChatGPT

There's a search box where you can discover GPTs, as you can see in Figure 4-1. There are also different categories you can select from, such as Writing, Productivity, Lifestyle, and Programming.

You can use GPTs if you have a free account with ChatGPT. But there are message limits. If you want to have a subscription, it is $20 per month.

Using a GPT is easy. You will click the icon, and there will be a pop-up, which will provide details about the GPT. To use it, click "Start Chat." Then you can chat with the GPT. Your GPTs will also appear on the top left of ChatGPT.

However, if you want to build a GPT, you need to have a paid account. If so, you can go to this URL for the GPT Builder:

```
https://chatgpt.com/gpts/editor
```

Figure 4-2. *This is the dashboard for the GPT Builder, which allows for creating GPTs*

Figure 4-2 shows the dashboard for it. On the left side, you will create the GPT, and the right side will preview the output.

Let's create a GPT.

You will first write a prompt for the GPT you want to create. It can be a sentence or two. Here are some examples:

- Product research that gathers information on products, compares features, and provides reviews

- Event planner that suggests themes, creates mood boards, and finds vendors and venues

- Fashion consultant that offers style advice, generates outfit ideas, and finds the latest fashion trends

The GPT Builder will then suggest a name for the GPT, a profile picture, and some default prompts. Of course, you can add your own or you can change them later.

The GPT Builder will ask a few more questions to refine the GPT. As you do this, you will be able to preview it on the right side of the screen. When it is finished, you can click Create at the top right of the screen. A menu will pop up, and you will have the option to share the link with whom you want. Click the Save button to create the GPT. You can start using it in ChatGPT.

Also, there are other features you can add to the GPT. You do this by selecting the Configure tab on the right side of the screen. Here, you have the option to change the profile picture and the prompt suggestions. You can also change the "Instructions." This is the prompt for the GPT.

Then there is the Knowledge feature. This allows you to upload documents, such as PDFs. This is a way to add data to your GPT for more specialization.

You can also select for capabilities like browsing the Web, image generation, and the use of the Code Interpreter, which uses Python to solve problems.

Then there are "Actions." With these, you can make connections to APIs, say from Slack, Notion, Zapier, and so on.

But there are some things to consider:

- GPTs are only accessed from ChatGPT. Rather, to integrate this capability into another website, you will use the Assistants API.

- Currently, there is no monetization for your GPTs. But OpenAI has plans to allow this.

- The developers of GPTs do not have access to the user conversations.

- The Enterprise edition of ChatGPT allows for creating GPTs that are for internal purposes.

Pricing and Tokens

When you use either the Playground or Assistants API, you will pay a fee for using an LLM. This is based on the number of tokens.

With the competition among the LLM providers, the pricing has been steadily declining. This is likely to continue for some time.

A token represents a word, part of a word, or a character. This structure depends on the LLM. But with a token, a model can process text.

A good way to understand how tokens work is to use OpenAI's Tokenizer, which you can find here:

```
https://platform.openai.com/tokenizer
```

Figure 4-3. *This is the OpenAI Tokenizer*

Figure 4-3 shows the page. At the top, the input box is where you enter text. We entered:

Acquiring knowledge continuously invigorates the intellect and maintains curiosity! ✸

Below, the Tokenizer uses a coding scheme to indicate the different tokens. The word "acquiring" is made up of two tokens and "knowledge" is one, which includes a space. As for "invigorates," it has three tokens. The exclamation has one token. And the emoji is made up of three as well.

In terms of the pricing, you can find the latest details here:
https://openai.com/api/pricing/

The page shows a listing of the available models, ordered by the performance. As of this writing, the top model is GPT-4o. You will pay $5 for each one million tokens. This is for the prompt. The response from the LLM, on the other hand, is $15 per one million tokens.

There are different flavors for the use of GPT-4o. For example, there is the Batch API. This means that the tokens are collected, and the LLM's responses are processed within a day or so. This comes with a 50% discount to the pricing. The Batch API is generally when you have high volumes for your application.

GPT-4o is multimodal, which means that it allows for recognizing and generating images. The pricing is also different—and more complex. Because of this, OpenAI provides a calculator. It's based on the width, height, and resolution of the image.

The next novel in performance is the GPT-4o model. It is much smaller—although it is still powerful. In fact, it is very useful for developing proof of concepts.

The GPT-4o model is much less expensive. The fee is 15 cents per one million tokens for the prompt and 60 cents per one million for the LLM responses. The Batch API is also at a 50% discount.

Another important type of model is for embeddings. This is for developing applications for search, topic modeling, classification, and clustering. There are different embedding models available. The pricing ranges from $0.020 to $0.10 per one million tokens.

Then there are models for fine-tuning. This is a more sophisticated way to customize an LLM. Fine-tuning is allowed for GPT-40-mini, GPT-3.5-turbo, davinci-002, and babbage-002. The pricing ranges from $0.30 to $12 per one million tokens.

Next, there is the pricing for the Assistants API. This is based on the tools you want to use. It is $0.03 per session for the Code Interpreter and $0.10 per GB per day for File Search.

Note   There is a rule of thumb: roughly 1,000 tokens are about 750 words.

OpenAI API

The OpenAI API is the system to build generative AI apps. It is not the same as the Assistants API.

But it's important to have a backgrounder in the OpenAI API. With it, you can build your own agents. This can provide more control over the application.

However, it will generally have much more code at least vs. using an agent framework.

For the purposes of this book, the focus will be on using frameworks—not natively creating agents.

In light of this, let's have a short demo of the OpenAI API. This program is a tweet generator. It asks for the user's API key once, then repeatedly prompts for topics. For each topic, it uses GPT-4o-mini to generate a one-sentence tweet with emojis. The program continues until the user chooses to quit.

We first need to install the OpenAI API:

```
pip install openai
```

Then we have the following:

```
import openai
import os
from getpass import getpass
```

First, we import the OpenAI library, which provides a Python interface to interact with OpenAI's API services.

Next, we bring in the os and getpass modules. These allow for turning the OpenAI API key into an environment variable. This is to secure it. You do not want anyone else to have access to the key—since they could charge your account for the fees.

This function will then ask to enter the API key:

```
def get_api_key():
    return getpass("Enter your OpenAI API key: ")
```

Then we have a function to generate the tweet:

```
def generate_tweet(client, topic):
    try:
        completion = client.chat.completions.create(
            model="gpt-4o-mini",
            messages=[
                {"role": "system", "content": "You are a social
                media expert skilled at creating engaging
                tweets."},
                {"role": "user", "content": f"Write a one-
                sentence tweet about {topic}. Include relevant
                emojis."}
            ]
        )
        return completion.choices[0].message.content
    except Exception as e:
        return f"An error occurred: {str(e)}"
```

The function takes two parameters: client (an instance of the OpenAI API client) and topic (the subject for the tweet). Inside a try-except block for error handling, it sends a request to the API using the chat completion endpoint. The request specifies the "gpt-4o-mini" model and includes two messages: a system message defining the AI's role as a social media expert and a user message requesting a one-sentence tweet about the given topic with emojis. If successful, it returns the generated tweet content from the API response. If an error occurs during this process, it catches the exception and returns an error message instead.

Finally, we have this function:

```python
def main():
    print("Welcome to the Tweet Generator!")

    # Get the API key once at the start
    api_key = get_api_key()
    client = openai.OpenAI(api_key=api_key)

    while True:
        topic = input("Enter a topic for your tweet (or 'quit'
                to exit): ")
        if topic.lower() == 'quit':
            break
        tweet = generate_tweet(client, topic)
        print("\nGenerated Tweet:")
        print(tweet)
        print("\n" + "-"*50 + "\n")
```

This starts by printing a welcome message, then securely obtains the API key from the user and initializes the OpenAI client. The function then enters a loop, repeatedly prompting the user for tweet topics. For each topic entered, it calls generate_tweet to create a tweet, prints the result, and adds a separator line. This cycle continues until the user types "quit," at which point the loop breaks and the program ends.

Assistants API

In November 2023, OpenAI announced its Assistants API. According to the company:

An assistant is a purpose-built AI that has specific instructions, leverages extra knowledge, and can call models and tools to perform tasks. The new Assistants API provides new capabilities such as Code Interpreter and Retrieval as well as function calling to handle a lot of the heavy lifting that you previously had to do yourself and enable you to build high-quality AI apps.[2]

Basically, this is a wrapper on the OpenAI API that allows for including agentic capabilities in generative AI apps. In fact, it's the underlying technology for GPTs.

Currently, the Assistants API is in beta. So when you read this chapter, some of the instructions will be different. But we'll have updates on the book's GitHub repository.

The Assistants system has two parts: the Playground and the API. Both operate on the same workflow, which you can find in Figure 4-4.

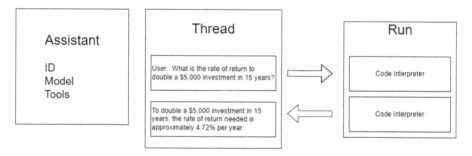

Figure 4-4. *This is the workflow for the OpenAI Assistants API*

First, we have the Assistant, which is the agent. It has a unique ID and is configured for an OpenAI LLM. You can also list the tools, and there are three: the Code Interpreter, File Search, and custom functions.

---

[2]https://openai.com/index/new-models-and-developer-products-announced-at-devday/

Next, there is a thread. This is where the user has a chat with the Assistant. The thread will store the messages, which are appended to the others. When this is done, all the messages will be sent to the LLM. This allows for memory. After all, an LLM does not have any state. For our thread, the message is about what the rate of return will be needed to double an investment. There is also the response, which is the answer from the LLM.

For this to be executed, there is the Run segment. It will take in the messages and use the tools. This is where there is orchestration. This process can go back and forth until the goals have been reached for the Assistant.

You will be charged for when you run an Assistant. This is based on the following price:

- Code Interpreter: This allows for running Python code, such as to handle computations, data analysis, and visualizations. OpenAI charges $0.03 per session.

- File Search: You can load and search external files. The fee is 10 cents per gigabyte of storage per day. OpenAI provides up to 1 GB free.

You will also be charged for when using a model. We'll discuss this later in the chapter.

Let's now take a look at the Playground.

Playground

With the Playground, you can experiment with the Assistants API. This is done in a low-code environment. To use the Playground, you can go to this URL:

```
https://platform.openai.com/assistants
```

Figure 4-5. *This is the introduction screen for the Playground of the Assistants API*

Figure 4-5 shows the screen. On the left side, you will see any existing Assistants you've created. Then on the right side, you will get the details for each of them. On the top right of the screen, you can click "Create" to create a new Assistant.

First, you will enter the name for the Assistant. It can be whatever you want.

Next, you will enter the instructions. This is essentially the "systems message" you would have for the OpenAI LLM. That is, it is the persona you want for the Assistant.

You will then select the model. The default is the most advanced one. But if you click the icon, you will get a drop-down of other models.

The next section is where you can specify the tools you want, which include File Search, Code Interpreter, and functions. For both File Search and Code Interpreter, you can add multiple files.

In the Model Configuration section, you can have the Assistant generate output in JSON format. You can also adjust for the following:

- Temperature: This is for the randomness or creativity for the model's output. It ranges from 0 to 2. The closer to 0, the more deterministic are the results.

- Top P: This controls the diversity of the LLM's output by considering only the most likely tokens whose cumulative probability reaches a specified threshold P. For example, if Top P is set to 0.9, the model will select from the smallest set of tokens whose combined probability is at least 90%. This method allows for more dynamic and contextually relevant responses.

You can select for the version of the API. However, it's common to use the default, which is the most recent.

You can also delete an Assistant or clone one.

Once you are finished with the Assistant, you can select Playground, at the top right of the screen.

You will then see the builder screen, which is in Figure 4-6.

Figure 4-6. *This is the builder for the Playground of the Assistants API*

The left side of the screen has the configuration options. Then in the middle, there is the thread. This shows the messages. To create a message, we enter text in the box below. There is also an option to add files and images.

Suppose we have the following message:

Assume the average return on my investments is 5% and I make $5,000 contributions each year. How long will it take to reach $100,000?

Figure 4-6 shows the output. The Assistant executes the Code Interpreter, which creates a Python program that calculates the investment amount. Then there is the result, which is that it will take 14 years to reach $100,000.

On the right side, you can see the logs for the Assistant. This can be helpful in seeing what is being executed. It can also allow for debugging.

On the top right of the thread, there are the number of tokens for the thread. You can also clear the thread and get a listing of the files being used.

71

Assistants API

To show how the Assistants API works, we'll create a program. It will calculate the return on investments.

The following imports the OpenAI class and creates an instance of it called client:

```
from openai import OpenAI
client = OpenAI()
```

Then we create the Assistant:

```
assistant = client.beta.assistants.create(
  name="Investing Bot",
  instructions="You are an expert in calculating investments.",
  tools=[{"type": "code_interpreter"}],
  model="gpt-4o-mini",
)
```

The "client.beta.assistants.create()" method is called to create a new Assistant. Note that this is for the beta.

The Assistant has several parameters:

- Name: This can be anything you want.

- Instructions: This is the system message for the Assistant.

- Tools: You can have more than one. But for this Assistant, we'll use the Code Interpreter.

- Model: We will use gpt-4o-mini.

We will create a thread and a message:

```
thread = client.beta.threads.create()

message = client.beta.threads.messages.create(
  thread_id=thread.id,
  role="user",
  content=" You have $10,000 to invest.  You want to double
  your money in 20 years.  What average return will you need
  to get?"
)
```

We will run the thread=ad:

```
run = client.beta.threads.runs.create(
  thread_id=thread.id,
  assistant_id=assistant.id,
  instructions="Provide detailed analysis."
)
```

The thread_id=thread.id specifies the conversation thread the Assistant will process. This ensures the Assistant has the context of the previous messages in the thread. Next, the assistant_id=assistant.id identifies which Assistant should be used for this run. This refers to the Assistant we created earlier.

Then instructions="Provide detailed analysis" gives additional instructions to the Assistant for this particular run.

It will take some time for this to be processed. This is why we create the following, which will notify us when it is completed:

```
import time

while run.status != "completed":
    run = client.beta.threads.runs.retrieve(
        thread_id=thread.id,
```

```
        run_id=run.id
    )
    print(run.status)
    time.sleep(5)
```

The import of time brings in Python's time module, which we'll use to add delays between status checks. The while loop will continue as long as run.status is not "completed." This means the loop will keep running until the Assistant finishes processing the request.

This code snippet retrieves all the messages from the thread:

```
messages = client.beta.threads.messages.list(
    thread_id=thread.id
)
```

The thread_id=thread.id specifies which thread's messages should be retrieved.

This code snippet processes and prints the messages from the thread, starting with the oldest message:

```
for thread_message in messages.data[::-1]:
    print(thread_message.content[0].text.value)
    print('\n')
```

The ([::-1]) is to allow for the reverse order. Then for each message, it accesses the text content using thread_message.content[0].text.value.

This function is designed to extract a percentage value from the text response:

```
def extract_rate_of_return(response):
    import re
    match = re.search(r'(\d+(\.\d+)?%)', response)
    if match:
        return match.group(1)
    return None
```

This function uses regular expressions to find and extract a percentage value from a given text response. It searches for a pattern of one or more digits, optionally followed by a decimal portion, and ending with a percent sign. If such a pattern is found, the function returns the matched percentage as a string; otherwise, it returns None.

We will then print out the response:

```
for thread_message in messages.data[::-1]:
    if thread_message.role == "assistant":
        rate_of_return = extract_rate_of_return(thread_message.
                        content[0].text.value)
        if rate_of_return:
            print(f"The average rate of return will need to be
            at least {rate_of_return}.")
        else:
            print("Could not extract the rate of return from
            the response.")
        break
```

This code iterates through the conversation messages in reverse order, looking for the Assistant's response. When it finds the Assistant's message, it attempts to extract the rate of return using the previously defined function. If successful, it prints the extracted rate; otherwise, it reports that extraction failed. The loop breaks after processing the first Assistant message, ensuring only the most recent response is analyzed.

As you use the API, you can create multiple Assistants. You can find details about them with this code snippet:

```
my_assistants = client.beta.assistants.list(
    order="desc",
    limit="20",
)
print(my_assistants.data)
```

This will show the Assistants in descending order and limit to 20.

Also, if you want to delete an Assistant, you can use the following:

```
response = client.beta.assistants.delete(my_assistants.
data[0].id)
print(response)
```

The my_assistants.data[0].id refers to the ID of the first Assistant in a list of Assistants.

Recent Advancements

The recent article by OpenAI on "Learning to Reason with LLMs"[3] introduces the o1 model, which is a new generation of LLM using reinforcement learning to achieve massive improvements in reasoning capability. For the most part, this is an example of how broad-based models are implementing agentic capabilities.

A key to the o1 model is that it can "think before it answers," thereby mimicking the way humans explain things. This is a major departure from traditional LLMs, which are based on the answers on the patterns they have learned from the training data.

However, with the o1 model, it engages in deep reasoning, breaking complex tasks into simpler ones so as to try to find the best response. It promises to do exceptionally well for domains that involve the most sophisticated problem-solving, such as competitive programming, mathematical reasoning, and science.

The article boasts about the achievements of the o1 model. About its performance on a variety of challenging benchmarks, o1 ranked in the 89th percentile on Codeforces, the competitive programming platform, and placed among the top 500 students in the USA Math Olympiad

---

[3] https://openai.com/index/learning-to-reason-with-llms

qualifier. On expert-level knowledge tasks such as the GPQA benchmark, o1 outperformed human PhD-level performance in physics, biology, and chemistry.

Table 4-1. *Performance of OpenAI o1 and Other Models in Various Competitions[4]*

Dataset	Metric	gpt-4o	o1-preview	o1
Competition Math AIME (2024)	cons@64	13.4	56.7	83.3
	pass@1	9.3	44.6	74.4
Competition Code CodeForces	Elo	808	1,258	1,673
	Percentile	11.0	62.0	89.0
GPQA Diamond	cons@64	56.1	78.3	78.0
	pass@1	50.6	73.3	77.3
Biology	cons@64	63.2	73.7	68.4
	pass@1	61.6	65.9	69.2
Chemistry	cons@64	43.0	60.2	65.6
	pass@1	40.2	59.9	64.7
Physics	cons@64	68.6	89.5	94.2
	pass@1	59.5	89.4	92.8
MATH	pass@1	60.3	85.5	94.8
MMLU	pass@1	88.0	90.8	92.3
MMMU (val)	pass@1	69.1	n/a	78.1
MathVista (testmini)	pass@1	63.8	n/a	73.2

---

[4] `https://openai.com/index/learning-to-reason-with-llms/#:~:text=their%20daily%20work.-,Appendix%20A,-Dataset`

So what is the main technical innovation of o1? It's about the use of reinforcement learning. This is a type of machine learning where an agent learns to make decisions by interacting with an environment and receiving feedback in the form of rewards or penalties. The agent's goal is to maximize cumulative rewards over time by optimizing its actions based on past experiences. So for o1, the model is trained on a trial-and-error basis. A chain-of-thought type of reasoning allows improvement either by seeing more data or by spending more "compute time" to arrive at more reflective and accurate answers.

OpenAI also explains the safety and alignment of the o1 model. By embedding human values within the chain of thought, the model becomes more effective in refusing an undesirable request or manipulative behavior. This has been the case when simulating jailbreaks and edge-case scenarios.

Overall, OpenAI claims that its o1 model offers unparalleled benefits compared to its predecessors in LLMs. As OpenAI continues to fine-tune this model, critical thinking and alignment to human values may give way to more application scenarios being discovered, hence enhancing its applicability in real life. Moreover, this shows that the company is investing heavily in agentic approaches.

In fact, here's what Aaron Levie—who is the CEO and founder of Box—has tweeted about o1:

> *At Box, we're seeing some very compelling results with OpenAI's new o1 model for improved reasoning on complicated enterprise data. For instance, it will follow multi-step logic rules in a contract to produce an answer, which we haven't seen before. This is pretty groundbreaking.*[5]

---

[5] https://x.com/levie/status/1835537106195918918?s=43&t=cS1w1V Zsy-iY3t91NeeUSw

Conclusion

By setting up an OpenAI account and obtaining an API key, users can start exploring the potential of powerful AI tools. GPTs allow for customization, enabling users to create personalized AI Assistants tailored to specific needs, whether it's for productivity, lifestyle enhancements, or business applications. Meanwhile, the Assistants API offers a more comprehensive solution for integrating AI capabilities into various applications, providing tools like the Code Interpreter and File Search to handle complex operations and data management.

The flexibility and power of OpenAI's platforms, combined with their user-friendly interfaces, make it easier than ever for developers and businesses to harness the potential of AI.

CHAPTER 5

Developing Agents

In the next part of this book, we will dive into the exciting world of frameworks for building AI agents, such as CrewAI, LangGraph, AutoGen, and Haystack. These frameworks are powerful tools that enable developers to create sophisticated, autonomous agents capable of performing a wide range of tasks. However, before we explore these frameworks in detail, it's crucial to establish a strong foundation by understanding the development resources and approaches that will guide our journey. This background knowledge will ensure that you're well prepared to tackle the more advanced concepts and techniques that lie ahead.

We'll begin by examining the various tools, libraries, and platforms available for AI agent development. From APIs to cloud services, these resources play a pivotal role in streamlining the development process, allowing you to focus on the logic and behavior of your agents rather than getting bogged down in technical details. We'll also discuss different development approaches, such as fine-tuning models and integrating external knowledge sources, which are essential for creating robust and effective AI agents.

To complement your learning experience, this book is accompanied by a GitHub repository, which you can access at `https://github.com/ttaulli/agents-book`. The repository contains code examples, sample projects, and additional resources that will help you apply the concepts discussed in this book. As you work through the chapters, you'll be able to follow along with the code and experiment with your own AI agent projects.

© Tom Taulli, Gaurav Deshmukh 2025
T. Taulli and G. Deshmukh, *Building Generative AI Agents*,
https://doi.org/10.1007/979-8-8688-1134-0_5

Jupyter Notebook, VS Code, and Google Colab

The choice of development environment can make a significant difference in how efficiently you can experiment, prototype, and deploy your AI models and agents. Jupyter Notebook, VS Code, and Google Colab are popular and versatile tools available. Here's a look.

Jupyter Notebook

Jupyter Notebook is a versatile web-based application that enables the creation of documents containing live code, visualizations, equations, and descriptive text, all within a single interface. This interactive environment is particularly favored by data scientists and AI practitioners for its ability to seamlessly combine code execution with rich media support. Its utility shines in exploratory data analysis and iterative model building, making it a preferred tool for developing and documenting AI projects. Additionally, the ease of sharing notebooks enhances collaboration and transparency. Jupyter Notebook allows users to execute code interactively with results displayed inline; supports various multimedia content like images, videos, and mathematical expressions; integrates effortlessly with popular data science libraries; and offers export options to formats like HTML and PDF for convenient sharing.

Visual Studio Code (VS Code)

Visual Studio Code (VS Code) is a powerful and free code editor developed by Microsoft, widely recognized for its flexibility and capability in developing generative AI applications. Its strength lies in supporting a broad range of programming languages while offering extensive customization through a variety of extensions. VS Code provides the functionality of a full-fledged integrated development environment (IDE), with features such as debugging, Git integration, and an integrated terminal. The rich ecosystem of extensions allows developers to tailor the editor to their specific needs, including seamless integration with Jupyter Notebooks and support for remote development in cloud environments. This makes VS Code a robust tool for AI development, with extensive language support, a vast marketplace of extensions, built-in terminal and debugging tools, and the ability to work on cloud-based or containerized projects directly from the editor.

Google Colab

Google Colab is a cloud-hosted Jupyter Notebook environment that distinguishes itself with its accessibility and computational resources. By offering free access to powerful GPUs and TPUs, Google Colab is particularly well suited for training and testing deep learning models, all without the need for any local setup. Running entirely in the browser, Colab integrates seamlessly with Google Drive, facilitating easy file storage and sharing. Its built-in collaboration features allow multiple users to work on the same notebook simultaneously, making it an excellent choice for collaborative AI projects. Google Colab also provides free, browser-based access to high-performance hardware, requires no installation, integrates directly with Google Drive, supports real-time collaborative editing, and comes preloaded with popular AI libraries like TensorFlow and PyTorch, significantly reducing setup time for developers.

How to Use Jupyter Notebooks

Given the centrality of Jupyter Notebooks in these tools, understanding their functions is crucial. So, let's take a demo. After launching Jupyter Notebook, you'll be directed to the notebook dashboard in your web browser. In the top-right corner, click the "New" button. From the drop-down menu, select "Python 3" (or any other kernel you prefer, depending on the programming language you are using).

A new notebook will open in a new tab. The notebook is initially untitled, which you can see in Figure 5-1.

Figure 5-1. *This is the initial screen for a new Jupyter Notebook*

To rename your notebook, click the "Untitled" text at the top of the notebook.

Enter a new name for your notebook in the dialog box that appears.

Click "Rename" to save the new name.

Jupyter Notebooks are organized into cells, which can contain either code or text. Let's start with code cells:

- The default cell type is a code cell, where you can write Python code or code in other supported languages.

- To execute the code within a cell, you can press Shift+Enter or click the Run button in the toolbar.

Suppose you enter this code into three cells:

```
# Calculate the sum of the first 10 natural numbers
sum_of_numbers = sum(range(1, 11))
sum_of_numbers
```

After running the last cell, the output will be 55. You can see this in Figure 5-2.

Figure 5-2. *This shows code entered in the cells of a Jupyter Notebook*

You can also use Markdown. This adds explanatory text, documentation, or any other descriptive content alongside your code.

To create a Markdown cell, click the drop-down menu in the toolbar (which typically shows "Code") and select "Markdown." You can then enter the text as well as formatting for headings, lists, links, and so on. After writing your content, run the cell to render the formatted text.

This is the sample Markdown:

```
# Summary of Results
The code above calculates the sum of the first 10 natural
numbers.

### Key Points:
- The sum is computed using Python's built-in `sum` function.
- The `range(1, 11)` generates numbers from 1 to 10.
- The final result, as shown in the code output, is 55.
```

Running this Markdown, you will get the output in Figure 5-3.

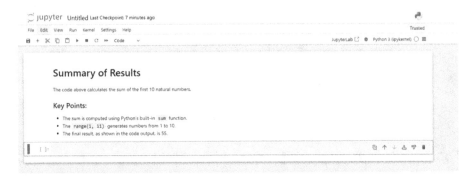

Figure 5-3. *This shows Markdown in Jupyter Notebook*

Of course, it's important to save your work regularly. You can do this by clicking the floppy disk icon in the toolbar or press Ctrl+S (or Cmd+S on Mac) to save your notebook. The notebook is saved in the .ipynb format, which you can open later from the Jupyter dashboard.

You can export your notebook to different formats for sharing or presentation purposes. Go to the File menu in the notebook and select "Download as" and choose the desired format (e.g., HTML, PDF, Markdown).

Google Colab

To use Google Colab, you will need a Google account. The platform is at colab.research.google.com.

The free tier, "Colab Free," provides access to standard computing resources, allowing users to run Jupyter Notebooks in the cloud without any cost. However, this tier comes with certain limitations, such as lower priority access to compute resources, a smaller GPU/TPU availability window, and session timeouts after periods of inactivity.

For users who require more robust computing power, Google Colab offers "Colab Pro" and "Colab Pro+" subscription plans. The Colab Pro plan, priced at $9.99 per month, enhances the user experience by providing access to faster GPUs and TPUs, longer session durations, and more memory. This plan is ideal for users needing reliable and faster processing for tasks such as machine learning model training. The Colab Pro+ plan, at $49.99 per month, offers even more resources, including the highest priority access to premium GPUs like the NVIDIA V100 and A100, extended sessions, and the ability to execute more intensive computational tasks with fewer interruptions. These paid tiers are designed to meet the needs of power users and professionals who rely on consistent and high-performance resources for their work.

But for the purposes of this book, we'll use the free version. Figure 5-4 shows the screen you'll see when you launch Colab.

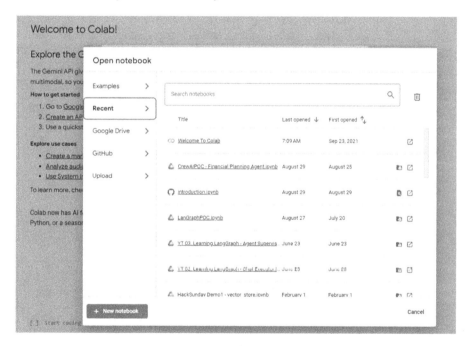

Figure 5-4. *This shows the initial screen when you launch Colab*

You'll see the recent files used, and there is a search box to locate them. Here are the other tabs:

- Examples: This has resources to get started with Colab and help with various features.

- Google Drive: You can import notebooks from your Google Drive.

- GitHub: You can import notebooks from GitHub.

- Upload: You can load files from your computer.

To get started, let's select the New Notebook button. Figure 5-5 shows what you get, which is a Jupyter Notebook as well as Colab-specific features.

Figure 5-5. *This is a notebook in Colab*

At the top, you can click Untitled0.ipynb and enter the name of the file you want.

You can then enter Code or Text into the cells. The text cells are equipped to process Markdown. Colab uses marked.js, which is similar but not quite identical to the Markdown used by Jupyter Notebooks or GitHub.

Streamlit, Gradio, and Jupyter Widgets

When creating AI agents, tools like Streamlit, Gradio, and Jupyter Widgets play an important role in making the development process more interactive, accessible, and user-friendly. These tools are designed to help developers build intuitive interfaces, enabling them to quickly prototype, test, and demonstrate AI models, making it easier to share their work with others, whether for research, collaboration, or deployment purposes.

- Streamlit: This is a powerful tool for creating custom web applications with minimal effort. It allows developers to turn Python scripts into interactive apps in just a few lines of code. This is particularly useful in generative AI, where visualizing the results of models, like text generation or image creation, is essential for understanding and refining the model's outputs. Streamlit's simplicity and flexibility enable developers to focus on their AI models rather than the complexities of web development, making it easier to iterate on ideas and share results with others, including nontechnical stakeholders.

- Gradio: This is another tool that simplifies the process of building user interfaces for machine learning models. It allows developers to create web-based demos for their generative AI models with just a few lines of code. Gradio is particularly valuable because it enables real-time interaction with models, allowing users to input data, adjust parameters, and immediately see the output. This interactivity is crucial for testing and refining generative models, as it provides immediate feedback and helps identify areas for improvement. Moreover, Gradio makes it easy to share these demos with others, facilitating collaboration and feedback from a broader audience.

- Jupyter Widgets: This extends the functionality of Jupyter Notebooks by allowing developers to add interactive elements such as sliders, buttons, and text boxes to their notebooks. This interactivity is invaluable in the generative AI development process, as it enables developers to tweak parameters and observe changes in model behavior in real time. For instance, when working with text generation models, Jupyter Widgets can allow users to adjust temperature settings or modify input prompts dynamically, offering a deeper understanding of how different inputs affect the model's outputs. This hands-on approach not only enhances experimentation but also aids in the educational process, making it easier to explain complex concepts to others.

Hugging Face

Hugging Face (`https://huggingface.co`) is a prominent platform and open source community that plays a key role in the artificial intelligence (AI) space, particularly in natural language processing (NLP) and generative AI. Since its founding in 2016, Hugging Face has become a go-to resource for developers and researchers, offering an array of tools, models, datasets, and libraries designed to simplify the creation and deployment of AI applications. Among its most significant offerings is the Transformers library, which includes pretrained models for various tasks such as text generation, translation, and summarization. These models, including well-known ones like GPT, BERT, and T5, can be easily integrated into projects, allowing developers to bypass the need for extensive training and computational overhead.

Hugging Face also features a comprehensive model hub, where users can access and share thousands of pretrained models, facilitating quick experimentation and deployment of AI solutions. This collaborative environment encourages developers to fine-tune existing models for their specific needs or contribute enhancements back to the community, speeding up the overall development process for generative AI applications.

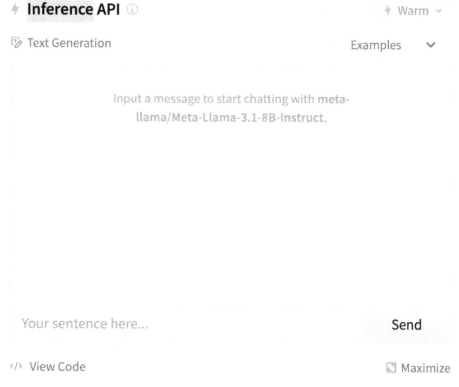

Figure 5-6. *Inference API for Meta-Llama-3.1-8B-Instruct Model*

Moreover, Hugging Face provides tools like the Inference API as shown in Figure 5-6, which simplifies the process of deploying models in production without requiring extensive infrastructure, and the datasets library, which offers easy access to a variety of datasets essential for training and evaluating models. By offering these resources, Hugging Face enables developers to focus more on innovative solutions and application-specific problems, rather than starting from scratch. For anyone working on generative AI applications, Hugging Face is a crucial resource that accelerates development, promotes collaboration, and provides easy access to cutting-edge models and tools.

Languages

There are various languages for the development of AI agents. R is often favored for statistical analysis and data visualization, making it useful in exploratory data analysis stages of AI projects. Java and C++ are known for their performance and scalability, which are crucial for deploying AI systems in production environments where speed and efficiency are paramount. Julia, with its high-performance capabilities and ease of use, is gaining traction for numerical and scientific computing tasks within AI.

Despite the availability of these languages, Python generally remains the preferred choice for developing generative AI applications. One of the main reasons for this preference is Python's extensive support for machine learning and deep learning libraries, such as TensorFlow, PyTorch, Keras, and Hugging Face's Transformers. These libraries offer prebuilt modules that simplify complex AI tasks, allowing developers to quickly prototype, experiment, and refine their models. Python's syntax is also user-friendly and concise, making it accessible to both beginners and experienced developers, which accelerates the learning curve and reduces development time.

Moreover, Python's strong community support and vast ecosystem of third-party packages make it highly adaptable to different AI-related tasks. Whether it's data preprocessing, model training, or deployment, Python provides tools and libraries that cover the entire AI development pipeline. Additionally, Python integrates well with other languages and systems, enabling seamless interaction with APIs, databases, and web frameworks, which is crucial for deploying generative AI models in real-world applications.

Python's versatility and widespread adoption also mean that there is a wealth of resources available, including tutorials, documentation, and forums, making it easier for developers to troubleshoot issues and find solutions to their problems. This community-driven support is invaluable, especially when working on cutting-edge AI projects that may require collaboration and knowledge sharing.

As for this book, we will be using Python.

Using LLMs (Large Language Models)

As a developer, you have multiple options for leveraging a large language model (LLM). They include the following.

Using an API from an LLM Provider

Many companies, like OpenAI, Anthropic, and Cohere, offer APIs that allow you to access their LLMs. You can integrate these APIs directly into your applications to generate text, answer questions, and more. Or you can use a framework like LangChain.

The pros of using an API are numerous. First, it offers ease of use, allowing you to access powerful LLMs without needing to understand the intricacies of model architecture or deployment. It's also highly scalable, as the provider handles traffic management, ensuring that your application can support a large number of requests without downtime. Additionally, these APIs often come with built-in security features, including data encryption and compliance with industry standards, which can save significant time and effort in securing your application.

However, there are some cons to consider. One of the main drawbacks is the lack of control over the model and its environment. You're limited to the configurations and capabilities provided by the API, which may not meet specific customization needs. Additionally, reliance on a third-party service means you're subject to their pricing models, which can become expensive as usage scales. There's also the concern of data privacy, as your data is being sent to external servers, which may not be ideal for sensitive or proprietary information. Lastly, latency can be an issue, as each request must travel to the provider's servers and back, potentially slowing down response times in latency-sensitive applications.

Using a Service like Ollama

There are several tools that make it easy to run LLMs locally, including on personal devices like laptops. Ollama is one such library, compatible with Windows, Mac, and Linux systems. With Ollama, you can load large models, such as those with 13 billion parameters, provided your machine has at least 16GB of RAM. The library supports a variety of open source models, including Mistral, Llama 2, and Gemma. Additionally, Ollama offers a REST API for running inference, allowing you to create applications powered by LLMs. It also features various Terminal and UI integrations that simplify the development of user-facing applications.

Then again, running LLMs locally with Ollama comes with several drawbacks. The resource intensity of these models can significantly impact your machine's performance, potentially slowing down other tasks or applications you're trying to run simultaneously. This is particularly noticeable on less powerful hardware or when trying to multitask. Another consideration is the manual effort required to keep models up to date. Unlike cloud-based solutions that often automatically use the latest versions, Ollama users need to actively manage and update their models to ensure they're using the most recent and improved versions. This can be time-consuming and may lead to using outdated models if not regularly maintained.

Using a Cloud Service like Azure, Google Cloud, or AWS

Major cloud providers offer AI services that include LLMs as part of their platforms. For example, Azure has OpenAI Service, Google Cloud offers Vertex AI, and AWS provides Amazon Bedrock and other AI services.

The pros of using a cloud service for LLMs are substantial, particularly in terms of scalability and integration. These platforms are built to handle enterprise-level workloads, so they can easily scale to meet the demands of high-traffic applications. Moreover, they offer a wide range of complementary services, such as data storage, security, analytics, and DevOps tools, which can be seamlessly integrated into your LLM workflows. This makes it easier to build complex, end-to-end solutions that go beyond just using the LLM. Cloud providers also offer strong security and compliance features, helping you manage data privacy and adhere to industry standards, which is particularly important in regulated industries.

However, there are cons as well. One of the primary drawbacks is cost. While cloud services offer pay-as-you-go pricing models, the costs can escalate quickly, especially with high-volume usage or when additional cloud services are required. Managing these costs effectively requires careful planning and monitoring. Additionally, while these platforms offer a high degree of control and customization, they also come with a steep learning curve. Developers need to be familiar with the cloud provider's ecosystem, which can be complex and require significant time to master. Another potential downside is the risk of vendor lock-in, where your application becomes tightly integrated with a specific cloud provider, making it challenging to migrate to another platform if needed. Finally, although these platforms offer strong security, there's always some degree of risk when storing and processing data on third-party servers, which may be a concern for organizations handling highly sensitive information.

Setting Up and Using Ollama

Ollama is available for Windows, Linux, and Mac systems. You can download it using this URL: ollama.com/download. The installation process is straightforward.

After this, go to your terminal and enter ollama. If you get resources about how to use the service, then you have successfully installed the program.

This URL shows the local models available on Ollama: ollama.com/library. Some of them you can download to a standard PC. But others could require systems with large amounts of RAM and have GPUs.

To use a model, this is the command:

```
ollama pull [name of the model]
```

Then to run it, you use this:

```
ollama run [name of the model]
```

You can then chat with the model:

```
from langchain_community.llms import Ollama
llm = Ollama(model="[name of the model]")
response = llm.invoke("What is 2 + 2 ?")
print(response)
```

However, it can be quite slow if you have a standard CPU machine. In fact, it can easily take a few minutes for the LLM to generate a response.

Because of this, you might consider using Ollama in Google Colab, since you will get access to more powerful GPUs and TPUs. There are different ways to do this, such as by using services like ngrok. This enables you to expose the local Ollama server running on Colab to the Internet, allowing your local machine to interact with it via a public URL, thereby leveraging cloud-based resources while interacting with LLMs as if they were running locally.

Using Ollama with Google Colab

An effective way to use Ollama is with Google Colab. You can leverage their GPUs and TPUs to run local models. When it comes to using proof of concepts, this can be free.

Here are the steps for using this approach. First, `InfuseAI/colab-xterm` is a utility to run terminals in Google Colab. We set this up with the following:

```
pip install colab-xterm
%load_ext colabxterm
%xterm
```

We run the local Ollama system by entering the following code in the shell that is instantiated by `colab-xterm`:

```
curl -fsSL https://ollama.com/install.sh | sh
ollama serve & ollama pull llama3 & ollama pull nomic-
embed-text
```

The curl utility downloads Ollama. And yes, it's that simple. You can then use a local model for your AI agents.

Why use a local model? When compared to using an API like OpenAI, there are several benefits. One is improved privacy in that all data resides on your machine. It might also be more cost-efficient because you do not pay any usage fees after the simple setup. Local models can also be much quicker with no dependency on Internet connectivity, which is a huge advantage in offline or real-time applications.

As for the rest of the chapter, we'll look at techniques for customizing LLMs.

Customizing LLMs

The true potential of LLMs often lies in their ability to be customized for specific use cases and domains. Customization allows LLMs to move beyond general-purpose applications and address specialized tasks with greater accuracy and efficiency.

Two key techniques have emerged to facilitate this customization process: fine-tuning and Retrieval-Augmented Generation (RAG). These approaches offer distinct methods for enhancing LLMs, enabling organizations to choose the best fit for their particular requirements and resources.

Fine-Tuning

This process for fine-tuning starts with a general LLM, which has been trained on a large and diverse corpus of text, and then refines it using a smaller, task-specific dataset. This typically involves several steps: first, you gather and preprocess the data that is relevant to the task or domain you're focusing on. Then, the LLM is trained on this dataset, adjusting its parameters to better understand and generate text that aligns with the specific context of your application. Finally, the fine-tuned model is evaluated and optimized to ensure it meets the desired performance criteria.

The pros of fine-tuning an LLM are important. It allows you to leverage the vast knowledge encoded in a general-purpose LLM while customizing it to excel at specific tasks. Some examples include answering customer service inquiries, generating legal documents, or writing technical content. This process can lead to more accurate and relevant outputs. It can help make the model more effective for specialized applications. Fine-tuning also reduces the need for extensive training from scratch. This saves time and computational resources.

However, there are some cons to consider. Fine-tuning requires access to a high-quality, domain-specific dataset. This may not always be readily available or easy to create. Furthermore, the process can be computationally expensive. This is especially the case when working with very large models. Fine-tuning also introduces the risk of overfitting. This is where the model becomes too specialized and loses its ability to generalize well to new, unseen data. Finally, fine-tuning can complicate the deployment and maintenance of the model. The reason is that it requires ongoing adjustments and updates to keep the model aligned with evolving data and use cases.

Fine-tuning methods generally fall into two categories: pretrained fine-tuning and advanced fine-tuning.

Pretrained fine-tuning leverages LLMs that have already been trained on extensive and diverse datasets. This method can be efficiently carried out using tools like Hugging Face Transformers, which offer a comprehensive set of resources for fine-tuning a variety of pretrained models and datasets. Another commonly used tool is PyTorch, a popular machine learning library that helps in training and fine-tuning models. What's more, many LLM providers, such as OpenAI, offer API services that allow users to fine-tune models without requiring deep technical expertise.

On the other hand, advanced fine-tuning techniques demand more specialized knowledge. One such method is Low-Rank Adaptation (LoRA), which reduces computational requirements and saves memory by simplifying the model's update process. QLoRA is a variation of LoRA that uses lower precision, thus improving the efficiency and speed of fine-tuning for larger models. Another advanced approach is Reinforcement Learning from Human Feedback (RLHF), where models are trained based on human feedback to ensure that their outputs align with human preferences. This is the technique used in interactive systems like ChatGPT, which asks users for feedback to refine its responses. A simpler yet effective alternative to RLHF is Direct Preference Optimization (DPO), which fine-tunes models to match human preferences while maintaining or improving performance in tasks like sentiment control, summarization, and dialogue generation. Unlike RLHF, DPO is easier to implement and train while achieving competitive or superior results.

Retrieval-Augmented Generation (RAG)

RAG is an advanced technique that enhances the capabilities of an LLM by incorporating external knowledge sources during the generation process. Instead of relying solely on the information embedded within the pretrained model, RAG involves retrieving relevant data from an external database—usually a vector database—or knowledge base in response to a query. The process typically involves two main components: a retrieval model and a generation model. When a query is received, the retrieval model searches through a large dataset to find the most relevant documents or pieces of information. These retrieved documents are then fed into the generation model, which uses this external knowledge to produce a more informed and contextually accurate response.

The pros of RAG are considerable, particularly in scenarios where up-to-date or specialized knowledge is essential. By supplementing the LLM with real-time or domain-specific data, RAG can produce more accurate and relevant outputs, making it especially useful in applications like customer support, research, and technical writing. This approach also mitigates the issue of "hallucination," where an LLM generates plausible but incorrect information, by grounding the generation process in actual data sources. Additionally, RAG enables models to remain useful over time without needing frequent retraining, as the external knowledge base can be updated independently of the LLM.

However, there are also cons to consider with RAG. The complexity of the system increases because it involves integrating and maintaining both a retrieval mechanism and a generation model. This can lead to additional challenges in terms of infrastructure, especially when handling large-scale deployments. RAG also introduces potential latency, as retrieving relevant documents from an external source takes time, which could impact the response speed in real-time applications. Furthermore, the effectiveness of RAG heavily depends on the quality and relevance of the external data; if the retrieval process pulls in irrelevant or outdated information, it could negatively affect the quality of the generated output. Lastly, implementing RAG requires careful tuning and optimization to balance the retrieval and generation components, which can be resource-intensive and require specialized expertise.

Conclusion

As we conclude this chapter on developing AI agents, it's clear that the landscape of AI development is rich with possibilities, offering a variety of tools, frameworks, and approaches to build sophisticated agents. The journey from understanding the foundational resources, such as APIs, cloud services, and local deployment options, to exploring advanced customization techniques like fine-tuning and Retrieval-Augmented Generation (RAG) sets the stage for creating powerful, specialized AI applications.

CHAPTER 6

CrewAI

Based in São Paulo, Brazil, João Moura has worked as a software engineer for the past 20 years. He has had stints at startups like Clearbit, Urdog, and Toptal. Along the way, he has developed systems in languages like Ruby, JavaScript, TypeScript, Elixir, and Python. He also has a strong background in machine learning and AI.

In 2023, he launched CrewAI, a Python library that he posted on GitHub. You can find it here:

`https://github.com/crewAIInc/crewAI`

What sparked his interest was the concept of having "fully autonomous departments backed by AI."[1] CrewAI would focus on building systems that rely on role-playing.

It did not take long for the framework to catch on. According to João:

> I've been testing CrewAI personally over the past few weeks and the results I'm getting have been impressive. It's like witnessing a jigsaw puzzle fall into place, each agent contributing to a bigger picture.[2]

Currently, the framework has over 18,000 stars and 115 contributors. In this chapter, we'll get a backgrounder on it.

---

[1] https://www.linkedin.com/posts/joaomdmoura_fully-autonomous-departments-backed-by-ai-activity-7130285095780777985--4DP/
[2] https://www.linkedin.com/posts/joaomdmoura_fully-autonomous-departments-backed-by-ai-activity-7130285095780777985--4DP/

T. Taulli and G. Deshmukh, *Building Generative AI Agents*,
https://doi.org/10.1007/979-8-8688-1134-0_6

The Basics

For AI agent frameworks, CrewAI is one of the easier ones. It is well thought out and intuitive. But it is still quite powerful.

Another key advantage is that CrewAI is built on the LangChain framework. This means that it benefits from its many integrations and functions. There is also the connection with LangSmith, which allows for testing and monitoring of generative AI apps.

As for CrewAI, it has various core concepts, such as agents, tasks, tools, processes, crews, and memory. We'll take a look at each of these.

Agents

An agent in CrewAI is an autonomous entity designed to execute tasks, make decisions, and interact with other agents within a collaborative team environment. Each agent acts as a specialized team member, equipped with specific skills that align with the team's overall objectives. For example, an agent could be a Data Analyst, Content Creator, or Technical Support specialist, each role crafted to support the team's goals effectively.

Here's sample code for an agent:

```
agent = Agent(
  role='Content Creator',
  goal='Develop engaging marketing content',
  backstory="""You're a content creator at a digital
  marketing agency.
  Your main responsibility is to create compelling content that
  resonates with the target audience and drives engagement.
  You're currently working on a campaign to promote a new
  product launch across social media platforms.""",
```

```
tools=[content_tool1, content_tool2],  # Optional, defaults
to an empty list
llm=my_llm,  # Optional
function_calling_llm=my_llm,  # Optional
max_iter=15,  # Optional
verbose=True,  # Optional
allow_delegation=True,  # Optional
callbacks=[callback1, callback2],  # Optional
)
```

The behavior and capabilities of an agent are defined by several key attributes:

> role: This defines the agent's specific function within the team. In this case, the agent is a "Content Creator," responsible for producing marketing content that aligns with the brand's messaging and goals.

> goal: This is the primary objective that guides the agent's actions. Here, the goal is to "Develop engaging marketing content," which drives the agent to focus on creating content that will capture the audience's attention and encourage interaction.

> backstory: Provides narrative context to the agent's role and objective. The backstory situates the agent as a content creator at a digital marketing agency, currently tasked with promoting a new product launch. This context helps the agent make decisions that are consistent with its role and responsibilities.

> tools: These are the specific capabilities or resources the agent can use to achieve its goal.

llm: The language model that powers the agent's ability to generate and understand text. my_llm represents the specific instance of a model like GPT-4, which the agent uses to produce content and make decisions.

function_calling_llm: A specialized language model responsible for handling the use of tools or functions by the agent. If specified, this model takes over the function-calling duties, ensuring that the agent can effectively utilize its tools to produce content.

max_iter: This sets a limit on how many times the agent can iterate on a task before finalizing its output. In this case, the agent has up to 15 iterations to refine and perfect its content before presenting the final version.

verbose: When enabled, this attribute ensures that the agent provides detailed feedback or logs during its operation. This is useful for tracking the agent's content creation process and understanding how it reaches its conclusions.

allow_delegation: Determines whether the agent can pass tasks or queries to other agents. With allow_delegation=True, the agent can delegate certain tasks to other specialized agents if it believes they are better suited for the task, ensuring efficient content creation.

callbacks: These are functions that are triggered at specific points during the agent's operation. This allows to call methods or functions to perform notifications or actions based on the outcome of the agent's operation.

Tasks

Tasks are distinct assignments managed by agents, designed with detailed instructions to ensure their successful execution. Each task provides everything an agent needs to complete its work, including descriptions, designated roles, required tools, and additional resources. This comprehensive setup allows agents to tackle tasks of varying complexity with precision.

In CrewAI, tasks can also be collaborative, often involving multiple agents working together. The properties of these tasks are configured to facilitate effective collaboration, with the Crew's processes ensuring that teamwork is streamlined and efficient.

Here's a code for a task:

```
task = Task(
    description='Research and compile a list of best practices
    for cybersecurity in cloud environments',
    expected_output='A detailed report outlining the top 10
    best practices for securing cloud environments, including
    explanations and examples.',
    agent=cybersecurity_agent,
    tools=[cloud_security_tool, research_tool]
)
```

This is a breakdown of the attributes in the task structure, along with explanations for each:

> description: This attribute provides a brief yet comprehensive summary of what the task entails. It outlines the main objective or purpose of the task, ensuring that the agent understands what needs to be done.

expected_output: This attribute defines the desired result or outcome of the task. It offers a detailed outline of what successful task completion should look like, ensuring that the agent knows what is expected.

agent: The agent is the specific entity or individual responsible for carrying out the task. This could be a designated agent or one determined through an internal process within the system.

Like with an agent, you can also set the attributes for max_iter, verbose, allow_delegation, and callbacks. But there are some others:

async_execution: This attribute allows the task to be run asynchronously, meaning it can be executed without halting other processes. This is useful when tasks are long-running or when multiple tasks need to proceed in parallel.

context: This attribute adds other tasks whose output can be provided as additional context for the execution of the task.

config: This attribute allows additional configuration details for further customization of the task.

output_json: This attribute enables the task to generate output in the form of a JSON object. This is useful for structured data that needs to be easily processed or transmitted, requiring an OpenAI client.

output_pydantic: Similar to the JSON output, this attribute allows the task to produce output as a Pydantic model object, which is especially useful for tasks that need to adhere to specific data models. This also requires an OpenAI client.

output_file: This attribute enables the task's result to be saved to a file. If combined with JSON or Pydantic output, it dictates how and where the output is stored, making it accessible for future use or review. Only one output format can be set.

human_input: Specifies whether the task requires human feedback upon completion, making it useful for tasks that need human oversight. By default, this is set to False.

callback: Call methods or functions to perform notifications or actions based on the task's operation output.

Tools

In CrewAI, a tool is a capability or function that agents use to perform a wide range of tasks, ranging from simple searches to complex interactions, all while supporting seamless teamwork. These tools are essential components that enhance agents' ability to collaborate effectively, drawing from options within the CrewAI Toolkit and LangChain Tools. These tools are designed with specific tasks in mind, such as web searching, data analysis, content generation, and agent collaboration, making them highly useful and versatile.

Integrated directly into the agents' workflow, these tools significantly boost the agents' capabilities, enabling them to execute tasks more efficiently. Additionally, these tools offer a high degree of customizability, allowing developers to create custom tools or utilize existing ones to meet the specific needs of the agents. To ensure smooth operations, they are equipped with robust error-handling mechanisms, and they feature intelligent caching to optimize performance by reducing redundant tasks. This combination of utility, integration, customizability, and performance optimization makes tools a vital part of the CrewAI ecosystem. This book uses the CrewAI version 0.51.0.

The following are some tools from CrewAI:

- BrowserbaseLoadTool: A tool for interacting with web browsers and pulling data from them

- CodeDocsSearchTool: A search tool optimized for exploring code documentation and technical manuals

- CSVSearchTool: A specialized tool for searching through CSV files, designed to handle structured data efficiently

- DALL-E Tool: A tool that generates images using the DALL-E API

- DOCXSearchTool: A tool for searching within DOCX files, making it easy to process and retrieve information from Word documents

- PDFSearchTool: A tool specialized in searching PDF documents, useful for handling scanned or complex files

- ScrapeWebsiteTool: A tool for scraping entire websites, perfect for collecting comprehensive data from the Web

- XMLSearchTool: A tool designed for searching through XML files, tailored for structured data formats

Crews

A crew is a collaborative assembly of agents working together to complete a set of tasks. Each crew is responsible for defining the strategy for task execution, coordinating agent interactions, and managing the overall workflow to ensure efficient and effective operations.

This is sample code for a crew:

```
project_team = Crew(
    agents=[data_analyst, report_creator],
    tasks=[data_analysis_task, generate_report_task],
    process=Process.sequential,
    full_output=True,
    verbose=True,
)
```

We have attributes to set the agents to run, as well as the tasks. There is also an attribute for the process. This is the sequence for how to run these agents. We'll discuss this in more detail below.

Next, there is the full_output attribute. This controls whether the crew returns all task outputs or just the final result.

Here are some other attributes:

- share_crew: This option allows you to share the crew's execution data and outcomes with the CrewAI team. Sharing this information can help improve the library and aid in model training.

- output_log_file: If you want to save a complete log of the crew's output and execution, you can set this to true or provide a path to a specific file where the log should be saved.

- planning: When activated, this attribute enables the crew to engage in planning before each iteration. You can also set the LLM for this with the planning_llm attribute.

Processes

In CrewAI, processes function as the backbone for task management, much like how project managers organize and delegate work in human teams. These processes make sure tasks are distributed and executed effectively, all according to a set strategy.

These are the types of processes:

- Sequential Process: Tasks are carried out one after another, following a set order. This ensures that each task is completed in a specific sequence.

- Hierarchical Process: Tasks are managed within a structured hierarchy. A manager, either a language model (manager_llm) or a custom agent (manager_agent), is responsible for task delegation and oversight. This setup allows the manager to create and assign tasks as needed, ensuring they align with the team's overall goals.

- Consensual Process: This process is designed to enable agents to make collaborative decisions regarding task execution. It introduces a democratic approach to task management in CrewAI but is currently planned for future development and isn't available in the existing code.

The following is the code for a sequential process:

```
crew = Crew(
    agents=my_agents,
    tasks=my_tasks,
    process=Process.sequential
)
```

And here is the code for a hierarchical process:

```
crew = Crew(
    agents=my_agents,
    tasks=my_tasks,
    process=Process.hierarchical,
    manager_llm=ChatOpenAI(model="gpt-4")
)
```

Memory

The CrewAI framework includes an advanced memory system that greatly enhances AI agents' abilities to remember, reason, and learn from past experiences. This memory system is composed of several key elements, each designed to improve the agents' performance in specific ways:

- Short-Term Memory: This temporarily stores recent interactions and outcomes, allowing agents to access and apply relevant information during current tasks. This immediate recall is essential for maintaining continuity and relevance in ongoing operations.

- Long-Term Memory: This preserves valuable insights from past executions, enabling agents to build a knowledge base over time and refine their decision-making processes. This accumulated wisdom allows agents to draw on past experiences to inform future actions.

- Entity Memory: This is focused on capturing and organizing information about entities, such as people, places, and concepts, encountered during tasks. This enhances the agents' understanding and ability to navigate complex scenarios by providing them with detailed knowledge about the key elements they interact with.

- Contextual Memory: This integrates the functions of short-term, long-term, and entity memory to maintain a consistent context across multiple tasks or conversations. This ensures that agents provide coherent and contextually relevant responses, even as the scope of their interactions expands.

By combining these memory components, the CrewAI system equips agents with contextual awareness, the ability to accumulate and learn from experiences, and a deeper understanding of key entities. This comprehensive memory system makes the agents more effective and intelligent in their interactions, enabling them to handle complex tasks with greater sophistication and adaptability.

For the rest of the chapter, we'll illustrate these core concepts with several CrewAI applications.

Note   CrewAI+ is the enterprise edition of CrewAI. It allows for quick transformation of any crew into an API, allowing for integration with applications through hooks, REST, gRPC, and other methods. There are also templates and custom tools, although, as of this writing, CrewAI+ is in a private beta.

Financial Planning Agent

We'll create a budgeting agent that assists users by analyzing their income and expenses to create a detailed monthly budget, ensuring that all necessary costs are covered while optimizing savings.

This involves multi-agent collaboration. First, there is the Budgeting Agent. It assists users by analyzing their income and expenses to create a detailed monthly budget. Next, there is the Investment Advisor Agent, which provides tailored investment recommendations based on the user's financial goals and risk tolerance, aiming to help the user grow their wealth through informed decisions. Lastly, the Debt Management Agent focuses on helping users manage and reduce their debt by developing effective repayment strategies and offering advice on debt consolidation and interest rate negotiations.

Let's go through the code:

```
!pip install crewai
!pip install crewai crewai-tools
```

We first install the core CrewAI package, which includes the essential classes and functions required to create and manage multi-agent systems within Python. Then we install the package for the tools. These tools include specialized components for managing specific types of agents, connecting to external APIs, or implementing advanced features within the multi-agent framework.

115

We then import the following:

```
from crewai import Agent, Task, Crew
```

The next part of the code listing is for defining the different agents. Here's the one for the Budgeting Agent:

```
budgeting_agent = Agent(
    role="Budgeting Advisor",
    goal="Create a monthly budget to help users manage their
    income and expenses effectively.",
    backstory=(
        "You are an experienced financial advisor specializing
        in personal finance. "
        "Your goal is to help users allocate their income
        efficiently, ensuring they cover "
        "all necessary expenses while also saving for the
        future."
    ),
    allow_delegation=False,
    verbose=True
)
```

The Agent class is instantiated with several key parameters that establish the agent's role, goals, and behavior. In this case, the agent is given the role of a "Budgeting Advisor," with the goal of creating a monthly budget to help users manage their income and expenses effectively.

The backstory parameter provides context and narrative for the agent, describing it as an experienced financial advisor specializing in personal finance. This backstory helps the agent frame its responses and actions in a way that aligns with the expertise expected from a financial advisor. The allow_delegation=False parameter ensures that this agent will not delegate its tasks to other agents, maintaining responsibility for its specific duties. Lastly, verbose=True enables detailed logging of the agent's actions, providing insights into how the agent processes information and makes decisions. This setup allows the budgeting agent to function effectively within a multi-agent system, contributing to the overall goal of comprehensive personal financial management.

Then we have the Investment Advisor Agent:

```
investment_agent = Agent(
    role="Investment Advisor",
    goal="Recommend suitable investment options based on the
    user's financial goals and risk tolerance.",
    backstory=(
        "You are an investment expert with years of experience
        in the financial markets. "
        "Your goal is to help users grow their wealth by
        providing sound investment advice "
        "tailored to their risk tolerance and financial
        objectives."
    ),
    allow_delegation=False,
    verbose=True
)
```

The backstory parameter is crucial in shaping the agent's responses and actions, portraying it as an experienced investment expert with a deep understanding of financial markets. This narrative helps the agent to deliver advice that is not only relevant but also aligns with the expertise expected from a seasoned financial professional.

Finally, there is the creation of the Debt Management Agent:

```
debt_management_agent = Agent(
    role="Debt Management Specialist",
    goal="Help users manage and reduce their debt through
    effective strategies.",
```

```
    backstory=(
        "You specialize in helping individuals overcome debt by
        creating personalized repayment plans. "
        "Your focus is on reducing interest payments and
        improving the user's financial health."
    ),
    allow_delegation=False,
    verbose=True
)
```

The backstory parameter improves the agent's capabilities by framing
it as an expert who specializes in creating personalized repayment plans
aimed at reducing interest payments and improving the user's overall
financial health. This narrative helps the agent provide advice and
solutions that are tailored to the user's unique financial situation, ensuring
that the guidance is both relevant and actionable.

For the next set of code, we will establish the three different tasks for
the agents. We start with this one:

```
budgeting_task = Task(
    description=(
        "1. Analyze the user's income and expenses. Financial
        Data: {financialdata}\n"
        "2. Create a detailed monthly budget that includes
        essential expenses, savings, and discretionary
        spending.\n"
        "3. Provide tips for optimizing spending and increasing
        savings."
    ),
    expected_output="A comprehensive monthly budget with
    recommendations for optimizing spending and savings.",
    agent=budgeting_agent
)
```

The code snippet is for the budgeting task. This involves a comprehensive approach, starting with an analysis of the user's income and expenses. The description parameter outlines the key steps the agent should take: analyzing the user's financial situation; creating a detailed monthly budget that includes essential expenses, savings, and discretionary spending; and finally, providing practical tips for optimizing spending and increasing savings.

The expected_output parameter specifies what the task should produce—a comprehensive monthly budget complete with recommendations for improving financial management. By linking this task to the budgeting_agent, the program ensures that the agent's expertise in financial planning is directly applied to generate a personalized and actionable budget for the user.

There is then the investment task:

```
investment_task = Task(
    description=(
        "1. Assess the user's financial goals and risk
        tolerance.\n"
        "2. Recommend suitable investment options such as
        stocks, bonds, mutual funds, or ETFs.\n"
        "3. Provide a brief overview of each recommended
        investment's potential risks and returns."
    ),
    expected_output="A personalized investment plan with
    recommendations and risk assessments.",
    agent=investment_agent
)
```

The description parameter outlines the steps the agent should follow: first, assessing the user's financial goals and risk tolerance to understand their unique investment needs; second, recommending suitable investment options such as stocks, bonds, mutual funds, or ETFs that align with these goals; and third, providing a brief overview of the potential risks and returns associated with each recommended investment.

The expected_output parameter specifies that the task should result in a personalized investment plan, complete with recommendations and risk assessments, tailored to the user's financial profile. By linking this task to the investment_agent, the program leverages the agent's expertise in financial markets to deliver sound investment advice, helping users make informed decisions that align with their financial goals and risk preferences.

The third task is for debt management:

```
debt_management_task = Task(
    description=(
        "1. Analyze the user's current debts, including
        interest rates and balances.\n"
        "2. Develop a repayment plan that prioritizes high-
        interest debt and suggests strategies for paying off
        balances faster.\n"
        "3. Provide advice on consolidating debt or negotiating
        lower interest rates."
    ),
    expected_output="A debt management plan with actionable
    steps to reduce and eliminate debt.",
    agent=debt_management_agent
)
```

The description parameter outlines the key actions that the agent will take: first, analyzing the user's current debts, including details such as interest rates and balances, to gain a comprehensive understanding of their debt situation; second, developing a repayment plan that prioritizes paying off high-interest debt first, which is often the most financially beneficial strategy; and third, providing advice on consolidating debt or negotiating lower interest rates to further reduce the user's financial burden.

The expected_output parameter specifies that the task should produce a detailed debt management plan that includes actionable steps for reducing and eventually eliminating debt. By assigning this task to the debt_management_agent, the program ensures that the agent's specialized knowledge in debt management is applied to create a personalized and practical plan for the user.

We will then need to connect the agents with the various tasks. We do this with the following code:

```
crew = Crew(
    agents=[budgeting_agent, investment_agent, debt_
    management_agent],
    tasks=[budgeting_task, investment_task, debt_
    management_task],
    verbose=True  # Set to True for detailed logging or False
    to reduce output
)
```

This Crew object brings together the three previously defined agents and the tasks parameter links these agents to their respective tasks, ensuring that each agent performs the actions assigned to it, such as creating a budget, recommending investments, and developing a debt management plan.

We will then run the Crew with this:

```
user_financial_data = dict({
    "financialdata": {
    "income": 5000,  # Monthly income in dollars
    "expenses": {
        "rent": 1500,
```

```
        "utilities": 300,
        "groceries": 400,
        "transportation": 200,
        "entertainment": 150,
        "other": 450
    },
    "debts": {
        "credit_card": {
            "balance": 2000,
            "interest_rate": 0.18   # 18% interest rate
        },
        "student_loan": {
            "balance": 15000,
            "interest_rate": 0.045   # 4.5% interest rate
        }
    },
    "savings_goal": 500 # Monthly savings goal in dollars
    }
})

# Now run the crew kickoff with the defined data
result = crew.kickoff(inputs=user_financial_data)

# Extract the raw text from the result
raw_result = result.raw

# Display the result as markdown
from IPython.display import Markdown
Markdown(raw_result)
```

The variable user_financial_data is defined to represent a user's financial situation, which includes various elements necessary for financial planning. The user's monthly income is set at $5,000. The expenses are broken down into categories such as rent ($1,500), utilities ($300), groceries ($400), transportation ($200), entertainment ($150), and other miscellaneous expenses totaling $450. Additionally, the user's debts are detailed, including a credit card balance of $2,000 with an 18% interest rate and a student loan balance of $15,000 with a 4.5% interest rate. The user also has a monthly savings goal of $500. This comprehensive financial data is then passed to the crew.kickoff function, which initiates the collaborative process involving multiple AI agents to generate financial advice. The result of this process is extracted as raw text, which is then displayed using Markdown for easy reading and interpretation, which you can see in Figure 6-1.

Figure 6-1. *This is the output for the CrewAI Agent*

Product Launch Orchestrator

We'll create a more advanced AI agent. We'll call it LaunchMaster. It's designed to streamline the complexities of a product launch, ensuring every aspect is meticulously managed for success. By leveraging specialized sub-agents, LaunchMaster handles crucial tasks such as conducting in-depth market research to identify target demographics and competitors, crafting engaging and persuasive content for various marketing channels, and executing targeted outreach to influencers and media outlets.

This program introduces some new concepts from the CrewAI framework. For example, we use new tools, such as SerperDevTool and ScrapeWebsiteTool. We also look at how to create a custom tool.

The agent will then allow for human input as well as provide for saving data to your computer.

Let's see how the program works.

We bring in the following libraries:

```
from crewai_tools import ScrapeWebsiteTool, SerperDevTool
from pydantic import BaseModel
```

ScrapeWebsiteTool is designed to scrape data from websites, enabling AI agents to gather and analyze web content dynamically. SerperDevTool searches the Internet. To use this, you will need to sign up for an API key at serper.dev.

From pydantic, we import BaseModel. It's a class used to define data models with validation rules in Python. This will provide the structure for the custom tool.

Then we create three agents:

```
market_researcher = Agent(
    role="Market Researcher",
    goal="Conduct thorough market research to identify target
    demographics and competitors.",
    tools=[search_tool, scrape_tool],
    verbose=True,
    backstory=(
        "Analytical and detail-oriented, you excel at gathering
        insights about the market, "
        "analyzing competitors, and identifying the best
        strategies to target the desired audience."
    )
)
```

```python
content_creator = Agent(
    role='Content Creator',
    goal="Develop engaging content for the product launch,
    including blogs, social media posts, and videos.",
    tools=[search_tool, scrape_tool],
    verbose=True,
    backstory=(
        "Creative and persuasive, you craft content that
        resonates with the audience, "
        "driving engagement and excitement for the product
        launch."
    )
)

pr_outreach_specialist = Agent(
    role="PR and Outreach Specialist",
    goal="Reach out to influencers, media outlets, and key
    opinion leaders to promote the product launch.",
    tools=[search_tool, scrape_tool],
    verbose=True,
    backstory=(
        "With strong networking skills, you connect with
        influencers and media outlets to ensure "
        "the product launch gains maximum visibility and
        coverage."
    )
)
```

The Market Researcher agent focuses on gathering and analyzing market data to identify target demographics and competitors, using tools like search and web scraping. The Content Creator agent develops engaging content, such as blogs and social media, to drive excitement and engagement for the product launch. The PR and Outreach Specialist agent targets influencers and media outlets to maximize the product launch's visibility and coverage, leveraging strong networking skills and the same tools for effective outreach.

This is the base class:

```
class MarketResearchData(BaseModel):
    target_demographics: str
    competitor_analysis: str
    key_findings: str
```

The MarketResearchData class is a structured data model designed to capture the output of the Market Researcher agent's analysis. This allows the agent to store and validate critical market research findings in a consistent format. The model includes fields for key audience segments, insights on market competitors, and key_findings, which highlights the most important conclusions drawn from the research.

Then we have a few tasks:

```
market_research_task = Task(
    description="Conduct market research for the {product_name}
    launch, focusing on target demographics and competitors.",
    expected_output="A detailed report on market research
    findings, including target demographics and competitor
    analysis.",
    human_input=True,
    output_json=MarketResearchData,
    output_file="market_research.json",
    agent=market_researcher
)
```

This task produces a detailed formatted output according to the MarketResearchData base model, ensuring consistency and accuracy. It also includes a human_input parameter, allowing for human feedback before finalizing the results. The output is saved as a JSON file ("market_research.json"), making it easy to share and integrate with other components of the launch strategy.

Next, there is this task:

```
content_creation_task = Task(
    description="Create content for the {product_name} launch,
    including blog posts, social media updates, and promotional
    videos.",
    expected_output="A collection of content pieces ready for
    publication.",
    human_input=True,
    async_execution=False,   # Change to synchronous
    output_file="content_plan.txt",
    agent=content_creator
)
```

This task instructs the Content Creator agent to develop various content pieces for a product launch. The task's goal is to produce a ready-to-publish collection of content that effectively promotes the product. There is also a parameter for human feedback. The completed content is saved in a text file ("content_plan.txt").

Finally, we have this task:

```
pr_outreach_task = Task(
    description="Contact influencers, media outlets, and key
    opinion leaders to promote the {product_name} launch.",
    expected_output="A report on outreach efforts, including
    responses from influencers and media coverage.",
    async_execution=False,   # Change to synchronous
    output_file="outreach_report.md",
    agent=pr_outreach_specialist
)
```

This directs the PR and Outreach Specialist agent to engage with influencers, media outlets, and key opinion leaders to promote a product launch. The task focuses on generating visibility and media coverage for the product. The expected output is a comprehensive report detailing the outreach efforts, including feedback and responses from the contacted parties. The final report is saved as a Markdown file ("outreach_report.md").

We create the crew for the agent:

```
product_launch_crew = Crew(
    agents=[market_researcher, content_creator, pr_outreach_
    specialist],
    tasks=[market_research_task, content_creation_task, pr_
    outreach_task],  # Ensure only one async task is at the end
    verbose=True
)
```

This crew consists of the Market Researcher, Content Creator, and PR and Outreach Specialist agents.

Then we run the crew:

```
launch_details = {
    'product_name': "SmartHome 360",
    'product_description': "A cutting-edge smart home system
    that integrates with all your devices.",
    'launch_date': "2024-10-01",
    'target_market': "Tech-savvy homeowners",
    'budget': 50000
}

result = product_launch_crew.kickoff(inputs=launch_details)
```

The launch_details dictionary contains key information about the product launch for the "SmartHome 360," a state-of-the-art smart home system that integrates seamlessly with various devices. It includes essential details such as the product name, a brief description, the launch date set for October 1, 2024, the target market of tech-savvy homeowners, and the allocated budget of $50,000. When these details are passed as inputs to the product_launch_crew via the kickoff method, the crew of AI agents begins executing their respective tasks—market research, content creation, and PR outreach—tailored specifically to these launch parameters.

The output will be extensive. For us, it was over 162 pages! The program accesses the Internet to gather data for the report and analysis.

Then we take this information and put it in a form that is more concise:

```
import json
from pprint import pprint
from IPython.display import Markdown
# Add this line to import the Markdown function

# Display the generated market_research.json file
with open('market_research.json') as f:
    data = json.load(f)
pprint(data)

# Display the generated content_plan.txt file
with open('content_plan.txt') as f:
    content = f.read()
print(content)

# Display the generated outreach_report.md file
Markdown("outreach_report.md")
```

First, the json module is imported to load and parse JSON data, while pprint from the pprint module is used to neatly format and print complex data structures.

The code then proceeds to open and load the market_research.json file, which contains data from the Market Researcher agent, and prints it using pprint for a clear and structured view. Next, the content_plan. txt file, generated by the Content Creator agent, is opened, read, and printed to display the planned content for the product launch. Finally, the outreach_report.md file, created by the PR and Outreach Specialist agent, is rendered using the Markdown function to present the outreach report in a well-formatted manner, making it easy to review the results of the agent's efforts. This sequence of steps ensures that all critical outputs from the AI agents are accessible and presented in a user-friendly format.

Customer Call Center Processing

We'll look at a more complex example using CrewAI. It will automate the task of a call center. For this, we'll use the hierarchical process.

A typical customer care center forms the service hub through which customers communicate, say with calls, emails, or chats. The calls from the customers are forwarded to agents according to the nature of the problem, which may be a billing query, technical support, or general inquiry, among others. The Agents provide direct solutions for the customer's problem or escalate problems to specialists. Complex issues will be tracked using a ticketing system and followed up to resolution. After the interaction, customers usually get asked for their feedback to improve the service quality further.

Figure 6-2 shows the hierarchy. Here are the agents:

- Customer Service Manager: Supervises several specialized agents for specific tasks

- Call Handling Agent: Manages tasks like call routing, query resolution, and escalations

- Technical Support Agent: Deals with troubleshooting, remote assistance, and service ticket creation

- Billing and Payments Agent: Handles invoice generation, payment processing, and dispute resolution

- Customer Feedback and Surveys Agent: Manages the survey distribution, feedback collection, and sentiment analysis

Figure 6-2. *The organizational chart of the customer call center*

Let's see how the program works.

We bring in the following libraries:

```
from langchain_openai import ChatOpenAI
from crewai import Crew, Process, Agent
```

As defined in the hierarchy above, we create the following four agents:

```
call_handling_agent = Agent(
    role="Call Handling Agent",
    goal="Manage and resolve customer inquiries via phone,
    including call routing and query resolution: {call_action_
    taken}",
    tools=[],
    allow_delegation=True,
    # This agent can delegate tasks if needed
    verbose=True,
    backstory=("Skilled in handling incoming calls, routing
    customers to the right departments, and resolving basic
    queries efficiently.")
)
```

The above code describes the functions of a Call Handling Agent in this system. Examples include call routing, response to simple queries, and escalation. This uses an argument that dynamically uses details of the action taken from the customer call itself {call_action_taken}. This agent also passes on tasks to other agents or departments for more complex problems.

We have the Tech Support Agent:

```
tech_support_agent = Agent(
    role="Technical Support Agent",
    goal="Troubleshoot and resolve technical issues reported by
    customers: {technical_action_taken}",
    tools=[],
```

```
    allow_delegation=True,  # This agent can delegate tasks to
    others or escalate when needed
    verbose=True,
    backstory=("Experienced in providing remote technical
    assistance, troubleshooting problems, and managing service
    tickets to ensure issues are resolved.")
)
```

The goal dynamically includes the action taken by the agent during resolution {technical_action_taken}. This agent has the capability to either delegate a task or escalate an issue to another agent (allow_delegation=True). With the description, the agent has broad experience with remote technical support, troubleshooting of all levels of problems, and service ticket management that covers all process stages.

Next, there is the Billing Agent:

```
billing_agent = Agent(
    role="Billing & Payments Agent",
    goal="Handle customer billing inquiries, process payments,
and resolve payment disputes: {billing_action_taken}",
    tools=[],
    allow_delegation=False,  # This agent does not
delegate tasks
    verbose=True,
    backstory=("Expert in managing billing-related issues,
ensuring accurate invoices, and processing payments quickly
while addressing customer disputes or issues.")
)
```

The goal dynamically includes the action taken regarding the customer's billing during this interaction {billing_action_taken}. Unlike other agents, this one cannot delegate tasks, as indicated by allow_delegation=False. The agent's description underlines the experience in handling all kinds of billing issues, such as correct invoicing, correct posting of payments, and sorting out any disputes or issues regarding billing.

Then there is the Feedback Agent:

```
feedback_agent = Agent(
    role="Customer Feedback & Surveys Agent",
    goal="Gather and analyze customer feedback through surveys
    and sentiment analysis: {customer_feedback}",
    tools=[],
    allow_delegation=False,   # This agent does not
    delegate tasks
    verbose=True,
backstory=("Specialized in collecting customer feedback,
conducting surveys, and analyzing sentiment data to improve
overall service quality.")
)
```

This agent analyzes the feedback by customers using {customer_feedback}. This agent does not allow for delegation. The backstory of an agent includes specialization for the collection of customer feedback, surveys, and sentiment data analysis to improve the quality of customer service.

The following code shows the tasks of various agents:

```
# Call Handling Task with context, agent, and callback
call_handling_task = Task(
    description="Handle incoming customer calls and resolve
    basic queries or escalate them if needed.",
    expected_output="A detailed report on calls handled,
    queries resolved, and escalations made.",
    human_input=True,
    output_json=CallHandlingData,
```

```
    output_file="call_handling_report.json",
    context={'priority': 'high', 'expected_resolution_time':
    '15 minutes'},  # Additional task context
    agent=call_handling_agent,
    callback=lambda result: print(f"Task completed by {call_
    handling_agent.role}: {result}"),  # Task callback
)

# Technical Support Task with context, agent, and callback
tech_support_task = Task(
    description="Troubleshoot technical issues reported by
    customers and resolve or escalate them as necessary as per
    {troubleshooting_steps}",
    expected_output="A report summarizing the technical issues
    handled and any open tickets.",
    human_input=True,
    output_json=TechSupportData,
    output_file="tech_support_report.json",
    context={'priority': 'medium', 'expected_resolution_time':
    '30 minutes'},
    agent=tech_support_agent,
    callback=lambda result: print(f"Task completed by {tech_
    support_agent.role}: {result}"),
)

# Billing & Payments Task with context, agent, and callback
billing_task = Task(
    description="Process customer invoices, handle payment
    issues, and resolve any disputes.",
    expected_output="A report detailing invoices processed,
    payments completed, and disputes resolved.",
    human_input=True,
```

```
    output_json=BillingData,
    output_file="billing_report.json",
    context={'priority': 'low', 'expected_resolution_time': '45
    minutes'},
    agent=billing_agent,
    callback=lambda result: print(f"Task completed by {billing_
    agent.role}: {result}"),
)

# Customer Feedback & Surveys Task with context, agent, and
callback

feedback_task = Task(
    description="Distribute customer satisfaction surveys and
    analyze the feedback for sentiment insights.",
    expected_output="A report summarizing survey results and
    customer sentiment analysis.",
    human_input=True,
    output_json=FeedbackData,
    output_file="feedback_report.json",
    context={'priority': 'low', 'expected_resolution_time': '60
    minutes'},
    agent=feedback_agent,
    callback=lambda result: print(f"Task completed by
{feedback_agent.role}: {result}"),
)
```

The Call Handling Agent handles customer calls, resolution of queries, and escalation of complex issues, which should be resolved within 15 minutes. The Technical Support Agent troubleshoots technical issues and prepares a report, which is expected to be done within 30 minutes. The Billing Agent handles invoice processing, payment problems, or disputes.

Overall, the priority is low, with a resolution window of about 45 minutes. The Customer Feedback Agent collects customer feedback via surveys. The priority of the task is also low, and the expected completion time is 60 minutes.

The `callback` attribute takes a function executed after the task has run, and a result is produced by logging a message that the agent has already finished the task. This is useful for real-time task execution tracking and management, maintaining a smooth workflow and proper recording of results.

For storage of the data related to tasks performed by the agents, we can create base classes as follows:

```
class CallHandlingData(BaseModel):
    call_summary: str
    resolved_queries: str
    escalated_issues: str

class TechSupportData(BaseModel):
    troubleshooting_steps: str
    resolved_issues: str
    open_tickets: str

class BillingData(BaseModel):
    invoice_details: str
    processed_payments: str
    resolved_disputes: str

class FeedbackData(BaseModel):
    survey_results: str
    customer_feedback: str
    sentiment_summary: str
```

The code snippet below is for creating a Crew that is composed of four agents:

```
customer_service_crew = Crew(
    agents=[call_handling_agent, tech_support_agent, billing_
    agent, feedback_agent],
    tasks=[call_handling_task, tech_support_task, billing_task,
    feedback_task],
    manager_llm=ChatOpenAI(temperature=0, model="gpt-4o"),
    # LLM managing overall task delegation
    process=Process.hierarchical,  # Hierarchical task delegation
    memory=True,  # Memory enabled for continuity in task execution
    planning=True,  # Enable planning to strategically allocate
    resources and manage tasks
)
```

Setting memory=True allows for storing information from prior tasks for better continuity in the performance of tasks. The attribute planning=True allows the system to make strategic plans through resource distribution.

The hierarchical process structure allows the crew to act like an organized and well-coordinated team in which every agent performs its role, and the system organizes the overall workflow and task distribution.

Next, we get input from the user:

```
customer_service_details = dict({
    'call_handling': {
        'queries_to_resolve': [
            'Account update request, need to update phone
            number',  # Customer had issues with updating
            their account
            'Complaint about slow service on previous calls'
        ],
```

```
          'call_action_taken': 'Account update request is
          completed. Call transferred to technical support'
}
          'escalations': {
              'technical_support': {
                  'issue': 'Product malfunction after
                  software update'
'technical_action_taken': 'noted the issues faced by the
customer and ticket created for further investigation, call
transferred to billing for resolution of overcharges'
              },
              'billing': {
                  'issue': 'Overcharged on most recent invoice',
                  'priority': 'High'
                  'billing_action_taken': 'The overcharges are
                  dropped and customer was sent to take survey
                  with end of the call'
              }
          }
      },
      'technical_support': {
          'troubleshooting_steps': [
              'Check for software compatibility issues',
              'Remote assistance to reinstall/update software'
          ],
      },
      'billing_payments': {
          'priority': 'High'
          'dispute_resolution_steps': [
```

```
            'Review the invoice for overcharges',
            'Calculate and issue refund if overcharge is
            confirmed'
        ]
    },
    'customer_feedback': {
        'feedback_sentiment': {
            'positive': '50%',  # The customer was satisfied
            with the billing resolution
            'neutral': '30%',  # Neutral on overall
            service quality
            'negative': '20%'  # Negative due to unresolved
            technical issue
        }
    }
}
)
```

Finally, we kick off our crew with the above user input:

```
task_result = customer_service_crew.kickoff(inputs=customer_
service_details)
```

And we print out the result of our execution at the end of crew task completion:

```
# Sample callback handling for task completion
def handle_task_completion(task_result):
    # Extract the raw text from the result
    raw_result = task_result.raw
# Display the result as markdown
    from IPython.display import Markdown
    print(f"Task completed:")
    Markdown(raw_result)
```

Retrieval-Augmented Generation (RAG)

CrewAI offers various classes for Retrieval-Augmented Generation (RAG). Here are some examples:

- CSVSearchTool: This allows agents to efficiently retrieve specific data from CSV files.

- DOCXSearchTool: This searches through DOCX documents, retrieving relevant sections of text based on the context of the query.

- PDFSearchTool: This allows for searching through PDF files, extracting and using relevant content for generating responses.

- WebsiteSearchTool: This searches through specified websites, pulling in pertinent information from web pages to enhance the generative process.

So let's see an example. We'll use WebsiteSearchTool. We start out with these imports:

```
from crewai import Agent, Task, Crew, Process
from crewai_tools import WebsiteSearchTool
```

We configure WebsiteSearchTool with a URL:

```
WebsiteSearchTool(website='https://en.wikipedia.org/wiki/Alan_
Turing')
```

This is a bio of Alan Turing.
We will create an Agent that performs a search:

```
search_agent = Agent(
    role='Website Researcher',
    goal='Search and extract relevant information from a
specific website.',
```

```
verbose=True,
memory=True,
backstory='You are an expert in searching websites for the
most relevant and up-to-date information.',
tools=[search_tool]
```

The agent is assigned the role of a "Website Researcher" with the clear goal of identifying and retrieving relevant data from the site. The agent's backstory establishes it as an expert in web research, emphasizing its capability to find the most relevant and current information. Finally, the agent is equipped with the necessary tools, in this case, the search_tool, to effectively carry out its search tasks.

We define the task for the agent:

```
search_task = Task(
    description=(
        "Use the provided website to find information on the
        topic '{topic}'. "
        "Make sure to gather all the relevant data available on
        this site."
    ),
    expected_output='A detailed summary of the information
    found on the website.',
    tools=[search_tool],
    agent=search_agent,
)
```

search_task instructs the agent to use the URL to gather information on a given topic, denoted by {topic}. The task description emphasizes the importance of collecting all relevant data available on the site. The expected output for this task is a detailed summary of the information found, which the agent will compile based on its findings. The task is equipped with the necessary search_tool to carry out the search process and is assigned to the search_agent, who will execute the task using its expertise in web research.

We will create the crew:

```
research_crew = Crew(
    agents=[search_agent],
    tasks=[search_task],
    process=Process.sequential
    # Executes tasks one after the other
)
```

The crew is configured to execute tasks sequentially. The inclusion of the search_agent ensures that a dedicated expert is in place to carry out the research, while the search_task provides clear instructions on what needs to be accomplished. By organizing the process in this way, the research crew efficiently handles the flow of tasks.

We kick off the crew with the input of "Artificial intelligence," which is for the topic, and then print out the results of the RAG:

```
result = research_crew.kickoff(inputs={'topic': 'Artificial
intelligence'})
```

```
print(result)
```

We then get an output that retrieves the relevant parts for the topic from the web page.

Connecting LLMs

Currently, the default LLM for CrewAI is GPT-4o. But the framework allows for many others.

Let's first look at how to use models from Hugging Face:

```
from langchain.llms import HuggingFaceHub
```

Then we create an instance of it:

```
llm = HuggingFaceHub(
    repo_id="HuggingFaceH4/zephyr-7b-beta",
    huggingfacehub_api_token="<HF_TOKEN_HERE>",
    task="text-generation",
)
```

Here's a look at the parameters:

- repo_id: This specifies the model you want to use from Hugging Face Hub.

- huggingfacehub_api_token: This is where you provide your API token for Hugging Face. You can get a free one at HuggingFace.co.

- task: This parameter specifies the type of task the model is designed for. In this case, "text-generation" indicates that the model will be used to generate text.

Another option is Cohere. For this, you will import this module:

```
from langchain_community.chat_models import ChatCohere
```

Next, you will set up an environment variable for the Cohere API key (which you can find at cohere.com):

```
os.environ["COHERE_API_KEY"] = "your-cohere-api-key"
```

You will then create an instance of the Cohere LLM:

```
llm = ChatCohere()
```

You can use the following for an Azure LLM:

```
from langchain_openai import AzureChatOpenAI

azure_llm = AzureChatOpenAI(
    azure_endpoint=os.environ.get("AZURE_OPENAI_ENDPOINT"),
    api_key=os.environ.get("AZURE_OPENAI_KEY")
)
```

There are several ways to use local LLMs. The preferred approach is with Ollama.

You will then use this import:

```
from langchain_ollama import ChatOllama
```

Then you will create an llm object with the following parameters:

```
llm = ChatOllama(
    model = "llama3.1",
    base_url = "http://localhost:11434")
```

This will connect the Ollama instance on port 11434.

Conclusion

Built on the robust LangChain framework, CrewAI simplifies the creation and management of multi-agent systems by focusing on role-playing scenarios, allowing agents to collaborate effectively to achieve complex goals. The framework's flexibility is evident in its ability to integrate with various tools, including web searching, data analysis, and content generation, making it highly versatile for different applications. With a growing community of contributors, CrewAI is poised to be a significant player in the AI development landscape.

CHAPTER 7

AutoGen

In August 2023, a group of researchers from Microsoft, Pennsylvania State University, the University of Washington, and Xidian University in China published a paper entitled "AutoGen: Enabling Next-Gen LLM Applications via Multi-Agent Conversation."[1] This paper introduced a new open source framework, AutoGen, for developers to build LLM-based applications that feature multiple agents working together.

These agents have proven effective in solving complex math problems, automating coding processes, and enhancing decision-making capabilities. By leveraging the strengths of multiple agents, AutoGen represents a significant advancement in LLM application development, offering a robust platform that can be tailored to a wide array of scenarios.

It has also seen a surge in popularity since its launch. Currently, AutoGen has over 30,000 stars on its GitHub repository, and there are more than 300 contributors.[2]

At present I am a lead principal software engineer, leading a team of ten people who are working on GenAI applications. We do use Agents for solving finance problems using AutoGen and Microsoft low-code/no-code copilot studio which also supports agentic workflows.

---

[1] https://arxiv.org/abs/2308.08155
[2] https://github.com/microsoft/autogen

"AutoGen gives more granular control to developers on defining the multi-agentic workflow and their customization," said Ravi Shankar Goli, who is a lead principal software engineer at Microsoft. He focuses primarily on generative AI. "The framework is easy to develop conversational agents when we need human-in-the-loop."

In this chapter, we'll take a look at the core components in AutoGen and how to create multi-agent systems.

Note   This chapter is based on AutoGen 0.2. However, there is expected to be a major update to the framework. But we'll have updates on our GitHub repo.

ConversableAgent

ConversableAgent is a specialized agent designed to manage conversations effectively. It handles input from users, processes it using predefined logic, and generates appropriate responses. The agent is built to understand and participate in natural language dialogues, making it useful in various chatbot and virtual assistant applications. It leverages predefined skills and contextual understanding to provide relevant and coherent interactions, ensuring a smooth and meaningful conversational experience for users.

Let's see how to use ConversableAgent. This will be a simple example of a conversation between two agents, Alice and Bob, where Alice is a friendly AI assistant, and Bob is a curious learner. The conversation will focus on a simple question-answer exchange, and we will print out the conversation details afterward.

We will have this setup:

```
pip install --upgrade pyautogen
from autogen import ConversableAgent
llm_config = {"model": "gpt-4o-mini"}
```

We start by installing the AutoGen library and then import
ConversableAgent. Then we set the LLM for gpt-4o-mini.

Next, we will set up the agents for Alice and Bob:

```
alice = ConversableAgent(
    name="alice",
    system_message="Your name is Alice, and you are a friendly
AI assistant ready to help with any questions.",
    llm_config=llm_config,
    human_input_mode="NEVER",
)

bob = ConversableAgent(
    name="bob",
    system_message="Your name is Bob, and you are a curious
learner who loves asking questions.",
    llm_config=llm_config,
    human_input_mode="NEVER",
)
```

The first agent, Alice, is set up as a friendly AI assistant. Her system
message defines her character, indicating that she is always ready
to help with any questions users may have. The configuration (llm_
config) specifies the language model she uses to generate responses,
and the human_input_mode is set to "NEVER," meaning Alice operates
autonomously without needing direct human input.

The second agent, Bob, is characterized as a curious learner who enjoys asking questions. Similar to Alice, Bob's behavior and responses are shaped by his system message, which highlights his inquisitive nature. Like Alice, Bob uses the same language model configuration and operates independently.

Then we use this to start the conversation:

```
chat_result = bob.initiate_chat(
    recipient=alice,
    message="Hi Alce! Can you tell me how photosynthesis
    works?",
    max_turns=2,
)
```

Bob starts the interaction by asking Alice a question about photosynthesis. The `initiate_chat` method is used to begin this dialogue, with Bob as the sender and Alice as the recipient. The initial message Bob sends is, "Hi Alice! Can you tell me how photosynthesis works?"

The `max_turns=2` parameter specifies that the conversation will continue for two exchanges (a turn from Bob and a response from Alice). This setup allows Bob to ask a question and receive an informative response from Alice, demonstrating how these agents can engage in a structured and coherent exchange, with Bob seeking knowledge and Alice providing it. The outcome of this interaction, including the conversation history and any generated responses, is stored in the chat_result variable, which can be further analyzed or displayed.

Reflection Agent

We'll look at the concept of reflection by creating a multi-agent system designed to create and refine tweets. Reflection, in this context, refers to the process where agents critically assess and improve upon their outputs based on feedback from other agents. This program will highlight how reflection is implemented by utilizing specialized agents—each with a distinct role, such as crafting the initial tweet, optimizing it for SEO, ensuring legal compliance, and providing final approval. Through this example, you'll learn how reflection enables these agents to iteratively refine their work, improving the quality and coherence of the final product.

We'll first need to load in the AutoGen library:

```
import autogen
```

Then we set the task:

```
task = '''
        Write an engaging tweet to promote a new AI tool
        designed for content creators.
        The tweet should be concise, include relevant hashtags,
        and be within the character limit for Twitter.
        '''
```

Basically, we want the tweet to be concise, effectively capturing the essence of the AI tool while fitting within Twitter's character limit. Additionally, it should include relevant hashtags to enhance its visibility and reach within the platform's ecosystem. This task serves as the foundational directive for the multi-agent system, guiding the subsequent steps where various agents refine and optimize the tweet to ensure it meets the specified criteria and maximizes its impact on social media.

We will set up various agents:

```
# Tweet Writer Agent
tweet_writer = autogen.AssistantAgent(
    name="TweetWriter",
    system_message="You are a tweet writer. You write concise
    and engaging tweets on given topics. "
```

```python
                        "Your tweets should be compelling, include
                        relevant hashtags, and be within the
                        character limit.",
    llm_config=llm_config,
)

# Content Optimizer Agent
content_optimizer = autogen.AssistantAgent(
    name="ContentOptimizer",
    system_message="You are a content optimizer. You refine the
    tweet to improve its clarity, engagement, and impact. "
                        "Your revisions should enhance the message
                        while keeping the tweet concise and within
                        the character limit.",
    llm_config=llm_config,
)

# SEO Reviewer Agent
seo_reviewer = autogen.AssistantAgent(
    name="SEOReviewer",
    llm_config=llm_config,
    system_message="You are an SEO reviewer. You optimize the
    tweet for search engines and social media algorithms. "
                        "Your suggestions should include relevant
                        keywords and hashtags that increase
                        visibility.",
)

# Legal Reviewer Agent
legal_reviewer = autogen.AssistantAgent(
    name="LegalReviewer",
    llm_config=llm_config,
```

```
    system_message="You are a legal reviewer. You ensure that
    the tweet is legally compliant and free from any potential
    legal issues. "
                        "Your review should be concise, ensuring that
                        the content adheres to legal standards.",
)

# Final Reviewer Agent (finalizes the tweet)
final_reviewer = autogen.AssistantAgent(
    name="FinalReviewer",
    llm_config=llm_config,
    system_message="You are the final reviewer. You aggregate
    all feedback and finalize the tweet, ensuring it is
    optimized, legally compliant, and engaging.",
)
```

In this multi-agent system, each agent plays a specialized role in the workflow, contributing to the creation and refinement of a high-quality tweet. The process begins with the Tweet Writer Agent, who is responsible for drafting the initial tweet. This agent sets the foundation for the tweet, ensuring that it captures the core message and appeals to the target audience.

Once the initial draft is created, the Content Optimizer Agent steps in. This agent refines the tweet by enhancing its clarity, engagement, and overall impact. The Content Optimizer's role is to polish the tweet, making sure that the language is sharp, the message is clear, and the content resonates well with the intended audience. This iterative process helps transform the initial draft into a more compelling piece of content, without sacrificing the brevity required for Twitter.

Next, the SEO Reviewer Agent evaluates the tweet from a search engine optimization perspective. This agent ensures that the tweet includes the right keywords and hashtags to maximize its visibility on social media platforms. The SEO Reviewer optimizes the content for algorithms, aiming to increase the tweet's reach and engagement by making it more discoverable to users who are searching for related topics.

The Legal Reviewer Agent then reviews the tweet to ensure it complies with legal standards. This agent is crucial for mitigating any potential legal risks, such as the use of copyrighted material, misleading statements, or any other content that could lead to legal issues.

Then the Final Reviewer Agent aggregates all the feedback from the previous agents and finalizes the tweet. This agent's role is to ensure that the tweet is not only legally compliant and SEO optimized but also engaging and polished. The Final Reviewer integrates the insights from all other agents, making any necessary adjustments before the tweet is approved for publication. This final step ensures that the tweet is of the highest quality, meeting all the necessary criteria for success on social media.

We will then have a function to generate a reflection message:

```
def reflection_message(recipient, messages, sender, config):
    return f'''Review the following content.
            \n\n {recipient.chat_messages_for_summary(sender)
            [-1]['content']}'''
```

This retrieves the latest message from the specified sender, which is then summarized and sent to the recipient for evaluation. This reflective process allows agents to critically assess each other's work, ensuring that feedback is given based on the most current output.

After this, we'll set up the nested review process:

```
review_chats = [
    {
        "recipient": content_optimizer,
        "message": reflection_message,
        "summary_method": "reflection_with_llm",
        "summary_args": {"summary_prompt":
            "Return review as a JSON object only:"
            "{'Reviewer': '', 'Review': ''}. Here Reviewer
```

```
should be your role",},
        "max_turns": 1
    },
    {
        "recipient": seo_reviewer,
        "message": reflection_message,
        "summary_method": "reflection_with_llm",
        "summary_args": {"summary_prompt":
            "Return review as a JSON object only:"
            "{'Reviewer': '', 'Review': ''}.",},
        "max_turns": 1
    },
    {
        "recipient": legal_reviewer,
        "message": reflection_message,
        "summary_method": "reflection_with_llm",
        "summary_args": {"summary_prompt":
            "Return review as a JSON object only:"
            "{'Reviewer': '', 'Review': ''}",},
        "max_turns": 1
    },
    {
        "recipient": final_reviewer,
        "message": "Aggregate feedback from all reviewers and
        finalize the tweet.",
        "max_turns": 1
    },
]
```

The review_chats list defines the sequence of reflection interactions among the various agents involved in refining the tweet. Each entry in this list specifies an agent (the recipient) who will receive a reflection message generated by the `reflection_message` function. The content optimizer, SEO reviewer, and legal reviewer are each tasked with reviewing the tweet's content, providing feedback within their areas of expertise. The summary method, `reflection_with_llm`, is used to summarize their feedback, and the agents are instructed to return their reviews as JSON objects, ensuring a structured and standardized format. The final entry in the list assigns the final reviewer the responsibility of aggregating all the feedback received from the previous agents and producing a finalized version of the tweet.

We will register the nested chats:

```
tweet_writer.register_nested_chats(
    review_chats,
    trigger=tweet_writer,
)
```

We will then initialize the chat and get the final tweet:

```
res = tweet_writer.generate_reply(messages=[{"content": task,
"role": "user"}])
res = final_reviewer.initiate_chat(
    recipient=tweet_writer,
    message=task,
    max_turns=2,
    summary_method="last_msg"
)
```

```
print(res.summary)
```

The process begins with the `tweet_writer` agent generating the initial tweet based on the provided task. The `generate_reply` function is used here to produce a response that serves as the draft tweet. Following this, the `final_reviewer` agent initiates a chat with the `tweet_writer` by calling the `initiate_chat` function. This function orchestrates a conversation between the agents, allowing the `final_reviewer` to gather and integrate all the feedback received during the earlier reflection phases. The `max_turns=2` parameter indicates that the chat can go back and forth up to two times to refine the tweet further. The `summary_method="last_msg"` ensures that the final output is based on the most recent message in the conversation, which should represent the finalized and refined tweet. The final result, encapsulated in `res.summary`, is printed out, showcasing the tweet that has been optimized and reviewed by all involved agents.

Figure 7-1 shows the output.

Figure 7-1. *This is the output for the Tweet Creator agent*

Tool Use

Our next program will be for an agent that processes a leave request for an employee. The program highlights the use of tools by integrating custom functions, which are registered as callable tools within the agents. By simulating a workflow where one agent summarizes a leave request and another agent makes an approval decision based on that summary, the program showcases how tools can be effectively utilized to break down complex tasks into modular, manageable components that agents can execute autonomously.

We'll first load in this library:

```
from typing import Annotated
```

Annotated allows you to add additional metadata to type hints, enabling you to provide more context or constraints for how a value should be used, which can be useful for documentation, validation, or tooling purposes in your code.

We then have a leave request to be processed:

```
leave_request_text = """
Employee: John Doe
Department: IT
Leave Type: Annual Leave
Leave Dates: 10/01/2024 - 10/05/2024
Total Days: 5
Reason: Vacation
Remaining Leave Balance: 10 days
Status: Pending Approval
"""
```

The variable leave_request_text is a multi-line string in Python that contains the details of a leave request. It includes information such as the employee's name, department, type of leave, leave dates, total days requested, reason for leave, remaining leave balance, and the current status of the request (pending approval).

We next have two tools:

```
def summarize_leave_request() -> Annotated[str, "Summary of the
leave request"]:
    """Summarizes the provided leave request."""
    return ("Summary: Employee John Doe from the IT department
    has requested 5 days "
            "of annual leave from 10/01/2024 to 10/05/2024 for
            vacation. "
            "Remaining leave balance is 10 days. Status is
            Pending Approval.")
```

```
def approve_or_reject_leave(summary: Annotated[str, "Summary of
the leave request"]) -> Annotated[str, "Approval or Rejection
decision"]:
    """Approves or rejects the leave request based on the
    summary."""
    if "Remaining leave balance is 10 days" in summary and "5
    days" in summary:
        return "Approved: The leave request is approved as it
        meets the company policy."
    else:
        return "Rejected: The leave request is rejected due to
        insufficient leave balance."
```

The summarize_leave_request function generates a summary of the leave request based on provided details, such as the employee's name, department, leave dates, and the status of the request.

The approve_or_reject_leave function takes the summary generated by the previous function as input and evaluates whether the leave request should be approved or rejected. It checks the leave balance against the days requested; if the balance is sufficient, the request is approved; otherwise, it is rejected. These functions demonstrate how the program automates the decision-making process by first summarizing relevant data and then applying business logic to reach a conclusion.

After this, we will create two agents:

```
# Leave Request Reviewer Agent
leave_request_reviewer = ConversableAgent(
    name="Leave Request Reviewer",
    system_message="You are responsible for reviewing the leave
    request. "
```

```
    "First, call summarize_leave_request() to get a summary of
    the leave request. "
    "Then, pass the summary to the Leave Approver for a
    decision.",
    llm_config=llm_config,
)

# Leave Approver Agent
leave_approver = ConversableAgent(
    name="Leave Approver",
    system_message="You are responsible for approving or
    rejecting the leave request. "
    "Wait for the summary from the Leave Request Reviewer
    and make a decision by calling approve_or_reject_
    leave(summary).",
    llm_config=llm_config,
)
```

The leave_request_reviewer agent is responsible for reviewing the
leave request. Its primary task is to call the summarize_leave_request
function to generate a summary of the leave details, which it then passes
on for further processing.

The leave_approver agent, on the other hand, is tasked with making
the final decision on the leave request. It waits for the summary provided
by the leave_request_reviewer and uses this summary to call the
approve_or_reject_leave function, which determines whether the leave
request should be approved or rejected. These agents illustrate how tasks
can be divided among different entities, each with specific responsibilities,
to streamline and automate a workflow.

We will then register the tools with the agents:

```
from autogen import register_function

# Register summarize_leave_request tool
register_function(
    summarize_leave_request,
    caller=leave_request_reviewer,
    executor=leave_request_reviewer,  # Set executor
    name="summarize_leave_request",
    description="Summarizes the leave request.",
)

# Register approve_or_reject_leave tool
register_function(
    approve_or_reject_leave,
    caller=leave_approver,
    executor=leave_approver,  # Set executor
    name="approve_or_reject_leave",
    description="Makes a decision to approve or reject the
    leave request based on the summary.",
)
```

For the summarize_leave_request function, it is registered with the leave_request_reviewer agent, making it callable and executable by this agent. Similarly, the approve_or_reject_leave function is registered with the leave_approver agent. By registering these functions as tools, the agents are equipped to perform their designated tasks within the workflow, enabling them to autonomously process and make decisions on the leave requests.

Finally, we have the code to run the multi-agent system:

```
summary_result = summarize_leave_request()
decision_result = approve_or_reject_leave(summary_result)

print(f"Summary: {summary_result}")
print(f"Decision: {decision_result}")
```

First, the `summarize_leave_request` function is called to generate a summary of the leave request. This summary is then passed to the `approve_or_reject_leave` function, which determines whether the leave should be approved or rejected based on the summarized information.

The results of these functions are stored in `summary_result` and `decision_result`, which are then printed out to display the summary of the leave request and the final decision. This manual simulation demonstrates how the entire process can be executed step by step, providing a clear understanding of how the agents and their associated tools interact to process a leave request.

And then we get this output:

> Summary: Employee John Doe from the IT department has requested 5 days of annual leave from 10/01/2024 to 10/05/2024 for vacation. Remaining leave balance is 10 days. Status is Pending Approval.
>
> Decision: Approved: The leave request is approved as it meets the company policy.

Group Chat

A group chat in AutoGen refers to a system where multiple agents collaborate to perform tasks through an automated chat framework. This feature enables a dynamic and interactive environment where these agents can communicate with each other to solve complex tasks collectively.

We'll look at a code example for Web Chat. It will be for the scenario of automating the process of resolving customer support tickets by engaging in a group chat with specialized agents. The agents will collaborate to understand the issue, suggest potential fixes, and determine the best course of action.

We first have this setup:

```
import autogen
llm_config = {"model": "gpt-4o-mini", "cache_seed": 42}
```

We import the library and then initialize it. We not only set the model but also the cache seed, which is for generating different outputs from the LLM while still using the cache.

We set up the User Proxy Agent, which is the customer service representative:

```
user_proxy = autogen.UserProxyAgent(
    name="Customer_Service_Rep",
    system_message="A human customer service representative.",
    code_execution_config={
        "last_n_messages": 2,
        "work_dir": "support_chat",
        "use_docker": False,
    },
    human_input_mode="TERMINATE",
)
```

The `system_message` attribute provides context, indicating that this agent represents a human role. The `code_execution_config` dictionary specifies that the agent will consider the last two messages in the conversation and will operate within the "support_chat" directory. We then indicate we will not use Docker.

Next, the `human_input_mode` is set to "TERMINATE," meaning the agent's interaction will end upon completing the task or when a certain condition is met.

We create agents for a technical support specialist and a
product expert:

```
tech_support = autogen.AssistantAgent(
    name="Tech_Support",
    system_message="An expert in technical troubleshooting.",
    llm_config=llm_config,
)

product_expert = autogen.AssistantAgent(
    name="Product_Expert",
    system_message="Knowledgeable in all product features and
    user issues.",
    llm_config=llm_config,
)
```

With the following, we set up the group chat:

```
groupchat = autogen.GroupChat(agents=[user_proxy, tech_support,
product_expert], messages=[], max_round=12)
manager = autogen.GroupChatManager(groupchat=groupchat, llm_
config=llm_config)
```

The GroupChat instance is initialized with these agents, and an
empty list of messages, indicating that the chat will begin with no
prior conversation history. The max_round parameter is set to 12,
limiting the conversation to 12 exchanges. This chat is managed by a
GroupChatManager, which coordinates the interactions among the agents.
Finally, we initialize the chat:

```
user_proxy.initiate_chat(
    manager, message="A customer reported that the software
    crashes during the export function. Investigate and provide
    a resolution."
)
```

Then, there will be an extensive back-and-forth with the different agents to resolve a customer service issue.

Web Search Agent

In the following example, we will show how to use `AssistantAgent` and `UserProxyAgent` for web retrieval.

`AssistantAgent` is an LLM-based agent that can write Python code. As for our example, we'll use it for scraping a website.

Next, `UserProxyAgent` is an agent that acts as a proxy for a user when executing code written by `AssistantAgent`. Depending on the setting of `human_input_mode`, `UserProxyAgent` can receive feedback from the user for the `AssistantAgent`. If the `human_input_mode` is "TERMINATE," then the `UserProxyAgent` will execute the code and return the execution result of success/failure. If there is user feedback, then `UserProxyAgent` passes it back to the `AssistantAgent`.

Here's the code:

```
assistant = autogen.AssistantAgent(
    name="assistant",
    llm_config=llm_config,
)

user_proxy = autogen.UserProxyAgent(
    name="user_proxy",
    human_input_mode="TERMINATE",
    max_consecutive_auto_reply=10,
    is_termination_msg=lambda x: x.get("content", "").rstrip().
    endswith("TERMINATE"),
    code_execution_config={
        "work_dir": "web",
        "use_docker": False,
```

```
    },  # Please set use_docker=True if docker is available to
    run the generated code. Using docker is safer than running
    the generated code directly.
    llm_config=llm_config,
    system_message="""Reply TERMINATE if the task has been solved
    at full satisfaction.
Otherwise, reply CONTINUE, or the reason why the task is not
solved yet.""",
)
```

Then we need to call the function initiate_chat() on the
UserProxyAgent to initiate the chat:

```
user_proxy.initiate_chat(
    assistant,
    message="""
What this article is about: https://pureai.com/
Articles/2024/03/01/autogen.aspx
""",
)
```

This will prompt the user at the end of each message whether the user
wants to provide feedback when the assistant agent raises a "TERMINATE"
signal. If the user simply presses Enter, the conversation will end
right away.

The advantage of using this functionality is that the web scraper
generated by UserProxyAgent will figure out the web page content and
structure and modify the code that needs to be scraped on the web page.
Figure 7-2 shows the output.

```
--------------------------------------------------------------------------------
assistant (to user_proxy):

Based on the HTML content of the webpage, I can see that the article is about the "AutoGen Framework" and its role
in advancing AI agents. Here is a summary of the article:

**Title:** Researchers Take AI Agents to the Next Level with the AutoGen Framework

**Summary:**
The article discusses the AutoGen Framework, an open-source project aimed at enhancing AI agents by enabling syste
ms where multiple agents can interact with each other and with humans. The framework is a collaborative effort fro
m Microsoft, Penn State University, and the University of Washington. It allows for the development of systems usi
ng multiple agents that can communicate and interact to solve complex tasks. The article highlights the customizab
le nature of AutoGen agents, their ability to integrate human participation, and the framework's capability to sim
plify the coordination and optimization of complex LLM workflows. Additionally, the article touches on the concept
of teachable agents, which can retain long-term memory and learn new skills over time.

**Key Points:**
- AutoGen is a software framework for developing LLM systems with multiple interacting agents.
- The framework is a collaborative research effort from Microsoft, Penn State University, and the University of Wa
shington.
- AutoGen agents can operate in various modes, combining LLMs, human inputs, and software tools.
- The framework simplifies the creation of multi-agent systems and optimizes LLM workflows.
- AutoGen includes a module for teachable agents, allowing them to retain long-term memory and learn new skills.

**Conclusion:**
The article concludes with insights from Dr. James McCaffrey from Microsoft Research and Ricky Loynd, the primary
architect of the AutoGen Teachable Agents module, emphasizing the potential of AutoGen to take AI systems to new l
evels of sophistication and capability.

TERMINATE

--------------------------------------------------------------------------------
Please give feedback to assistant. Press enter or type 'exit' to stop the conversation: thank you! exit
user_proxy (to assistant):

thank you! exit

--------------------------------------------------------------------------------
assistant (to user_proxy):

You're welcome! TERMINATE

--------------------------------------------------------------------------------
```

Figure 7-2. *This is the final output for the Web Search Agent*

Retrieval-Augmented Generation (RAG)

We will create a program that uses Retrieval-Augmented Generation (RAG). The scenario will be to assist entrepreneurs in developing effective business plans by combining an LLM with document retrieval. The program initializes two agents: an assistant agent, designed to act as an experienced business consultant, and a proxy retrieval agent, which fetches relevant business-related documents. The retrieval agent accesses specific resources, including business plan templates and guides from reliable sources, to ensure the assistant provides accurate, document-supported advice.

167

Here's the setup code:

```
from autogen.agentchat.contrib.retrieve_assistant_agent import
RetrieveAssistantAgent
from autogen.agentchat.contrib.retrieve_user_proxy_agent import
RetrieveUserProxyAgent
llm_config = {"model": "gpt-4o-mini", "timeout": 600, "cache_
seed": 42}
```

This begins by importing the required modules, including RetrieveAssistantAgent and RetrieveUserProxyAgent, which are responsible for handling an assistant agent and a proxy agent for document retrieval. The llm_config defines the configuration for the language model, specifying the use of the "gpt-4o-mini" model, setting a timeout of 600 seconds, and adding a "cache_seed" for reproducibility.

Then we initialize the assistant agent:

```
assistant = AssistantAgent(
    name="business_assistant",
    system_message="You are an experienced business consultant
    helping entrepreneurs.",
    llm_config=llm_config,
)
```

The system_message sets the assistant's role and tone for interactions. The llm_config is passed to the agent to specify the LLM and other configuration details.

Next, we use the Retrieve Proxy Agent to fetch business-related documents from the Internet:

```
ragproxyagent = RetrieveUserProxyAgent(
    name="business_ragproxyagent",
    human_input_mode="NEVER",
    max_consecutive_auto_reply=3,
```

```
    retrieve_config={
        "task": "business",
        "docs_path": [
            "https://www.sba.gov/business-guide/plan-your-
            business/write-your-business-plan",
            "https://www.score.org/resource/business-plan-
            template-startup-business",
            os.path.join(os.path.abspath(""), "..",
            "business_docs"),
        ],
    },
    code_execution_config=False,   # No code execution for this
    business task
)
```

The human_input_mode is set to not have human intervention and max_
consecutive_auto_reply to allow the agent to respond automatically up
to three times before requiring further action. The retrieve_config defines
the task as "business" and specifies a list of document paths, including
external URLs and a local path, from which the agent can retrieve
business-related information. This is information from the Small Business
Administration (SBA).

The code_execution_config is set to "False," indicating that no code
execution is necessary for this task, as the focus is purely on document
retrieval for business consulting purposes.

We specify the business question:

```
business_problem = (
    "How can I create an effective business plan for a small
    retail store, "
    "and what should I include in the financial projections?"
)
```

Finally, we initiate the retrieval and conversation with the assistant agent:

```
ragproxyagent.initiate_chat(assistant, problem=business_
problem, search_string="business plan, financial projections")
```

The initiate_chat method is called on the ragproxyagent, which prompts the retrieval agent to begin the process of assisting the assistant in providing a solution. The assistant will leverage the retrieved documents and resources to help answer the defined business_problem, which in this case involves creating an effective business plan and providing financial projections. The search_string parameter further narrows the focus of the retrieval by specifying relevant keywords like "business plan" and "financial projections," guiding the agent to locate appropriate resources on these topics.

When you run this, you will get an input box to enter a prompt. You can enter something like this:

For a business plan for a retail store, write an executive summary.

The RAG system will then search the document and provide a response.

Using Ollama

You can use a local LLM with AutoGen. A common approach for this is to use Ollama.

First, you will create a configuration list for the Code Llama model running locally, which will include the local server's base URL where the model is running ("http://localhost:11434/v1") and the API key:

```
config_list = [
  {
    "model": "codellama",
    "base_url": "http://localhost:11434/v1",
```

```
    "api_key": "ollama",
  }
]
```

Then we can create an assistant that uses the local model configuration:

```
assistant = AssistantAgent("assistant", llm_config={"config_
list": config_list})
```

We can then use a proxy agent with the code execution configurations:

```
user_proxy = UserProxyAgent("user_proxy", code_execution_
config={"work_dir": "coding", "use_docker": False})
```

You can then use the local model, such as by using a prompt:

```
user_proxy.initiate_chat(assistant, message="Write a Python
script to scrape the latest headlines from a news website
like BBC.")
```

AutoGen Studio

AutoGen Studio is a powerful tool designed to help you create generative AI agents with a low-code approach. You can access the AutoGen Studio at the following link:

```
https://autogen-studio.com
```

To begin, make sure you have Python 3.11 or a more recent version installed. You can easily install Python using Conda, which is available here:

```
https://anaconda.org/anaconda/conda
```

Next, since AutoGen Studio relies on OpenAI's LLMs, you'll need an OpenAI API key. If you're using Microsoft Azure, you can use an API key from there as well.

Let's now see how to set up your environment. On both Windows and macOS, open your terminal or command prompt and run the following command:

```
conda create -n autogenstudio python=3.11
```

Then activate the environment:

```
conda init
conda activate autogenstudio
```

You will set up your OpenAI API key. On macOS or Linux:

```
export OPENAI_API_KEY=XXXX
```

On Windows:

```
set OPENAI_API_KEY=XXXX
```

Next, we'll install the AutoGen Studio:

```
pip install autogenstudio
```

To launch it, you will do the following:

```
autogenstudio ui
```

After launching, copy the provided URL into your browser to access the UI. Figure 7-3 shows the initial screen.

Figure 7-3. *This is the initial screen for the AutoGen Studio*

At the top left of the screen, you have two options:

- Build: This is where you create your AI agents.

- Playground: This section allows you to test and experiment with your agents.

Let's first look at Build. On the left side of the screen, you have various menu items. One is Skills. This lets you create Python functions for specific tasks. You can use existing functions, like generating and saving images or PDFs, or create new ones. As you define these skills, the corresponding Python code is automatically generated and updated. You can even copy and paste this code into your own IDE if you prefer.

On the top right of the screen, you can select New Skill. This allows you to create a custom skill.

Figure 7-4 shows the screen for this.

Figure 7-4. This allows you to create a custom skill

You can provide a name for it and a description, which is a prompt that will tell the LLM what to do. You can then enter the LLM you want to use.

As you enter these parameters, the code on the left side will be updated.

So for the Name, I put tweet_creator, and then for Description, I entered "Create an interesting tweet." Then I clicked Save Skill.

The next menu option is Models. Yes, here you specify the LLM you want to use. AutoGen Studio supports several preconfigured models like GPT-4. Or, you can add a model that is not on the list. You do this by

174

selecting New Model and entering the information, such as the model name and the API key. You can then click Test Model to see if there is a connection. Figure 7-5 shows this for gpt-4o.

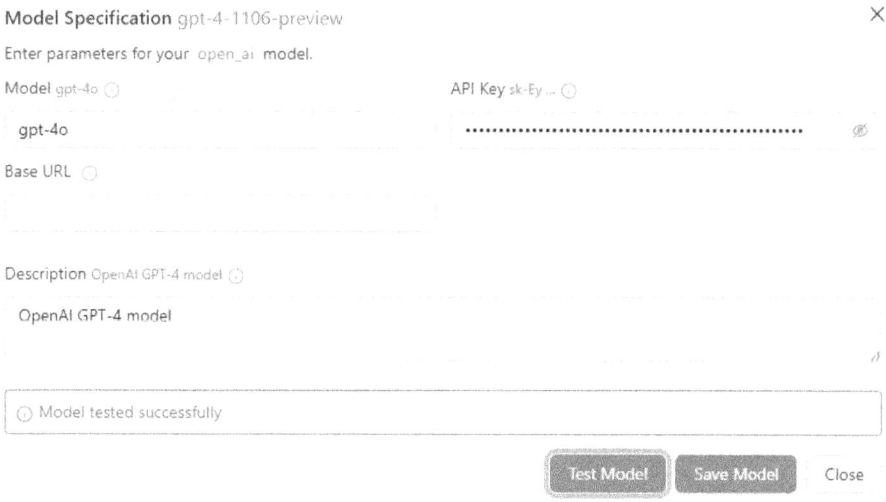

Figure 7-5. *This shows the selection of a new model*

Next, we have the Agents menu item. With this, we can configure an agent for your workflow. There are different options, such as an agent for planning and one for a language assistant. You can see these in Figure 7-6.

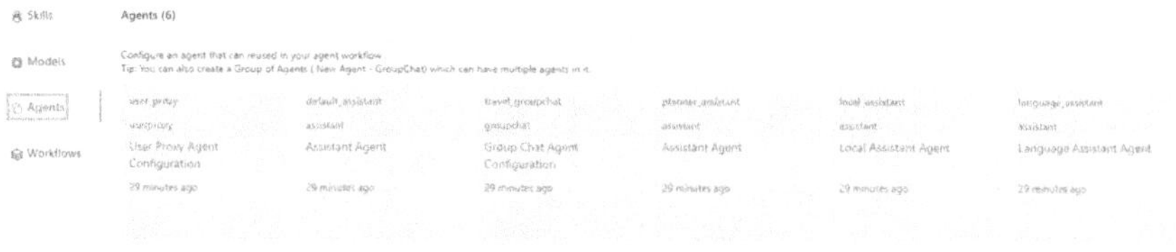

Figure 7-6. *These are default agents*

You can create a custom agent by selecting New Agent. You have different templates. They are

> User Proxy Agent: This represents a user and executes code.

> Assistant Agent: This is for planning and generating code to solve problems.

> GroupChat: This is to manage group chat interactions.

Finally, there is the menu item for Workflows. This is where you connect multiple agents to create complex, multi-agent interactions. For example, you could build a workflow for travel planning, where various agents communicate and collaborate.

With our Agent, we can test it out in the Playground. We will select New, which will create a session.

You will then select the type of workflow. I will choose the default. Then I will click Create.

Figure 7-7 shows the screen for this.

Figure 7-7. *This is a session for the Playground*

Then you can test it by writing a prompt or using a predefined one.

Conclusion

AutoGen offers a powerful framework for building multi-agent systems that can efficiently solve complex tasks across a wide range of applications. From generating engaging social media content to automating approval workflows, AutoGen enables developers to create highly specialized agents that work together to achieve precise outcomes. The modularity and flexibility of this platform allow for seamless integration of custom tools, enhanced collaboration among agents, and the ability to handle diverse scenarios. As the development of multi-agent systems continues to evolve, AutoGen stands out as a pioneering framework that brings robust functionality to developers.

CHAPTER 8

LangChain

LangChain is an open source framework designed to simplify the development of applications powered by LLMs. It addresses the complexities of working with LLMs by providing a comprehensive suite of tools, components, and prebuilt chains that enable developers to create sophisticated, context-aware AI applications efficiently. LangChain facilitates the seamless integration of LLMs with external data sources and computational resources. This allows for more dynamic and interactive AI experiences.

Understanding LangChain is important for developers who want to create AI agents within the LangGraph framework, which we will cover in the next chapter. LangGraph, which is built on top of LangChain, extends these foundational capabilities by incorporating advanced graph-based models that enhance the logical structure and decision-making processes of AI agents. It allows developers to create agents that can navigate complex decision trees, manage intricate state transitions, and optimize task execution in a more structured and efficient manner. Without a solid grasp of LangChain, developers may find it challenging to fully harness the power of LangGraph's advanced features.

This interdependence is also evident in frameworks like CrewAI, which we will cover in Chapter 6. This similarly builds on LangChain to offer specialized capabilities for AI agent development.

© Tom Taulli, Gaurav Deshmukh 2025
T. Taulli and G. Deshmukh, *Building Generative AI Agents*,
https://doi.org/10.1007/979-8-8688-1134-0_8

While LangChain itself provides robust tools for creating AI agents, enabling functionalities like prompt engineering, data integration, and multi-agent coordination, frameworks like LangGraph and CrewAI take these capabilities to the next level. They allow for more specialized and high-performance applications, making LangChain a foundational skill for any developer working in this space. Understanding the nuances of LangChain not only equips developers to build effective AI agents but also positions them to leverage more advanced frameworks that rely on its core principles.

LangChain is available for both Python and JavaScript. But for this book, we will focus on Python.

Background

Harrison Chase developed LangChain to address challenges commonly faced by AI developers when building LLM-based applications. The process of creating these applications is often complex, tedious, and time-consuming, requiring the integration of various components and prompts. LangChain was designed to simplify this process by allowing developers to "chain" multiple prompts and components together, streamlining development and reducing the time and effort needed to create powerful AI-driven applications. This ability to orchestrate different elements within an LLM application is a key feature, which is why LangChain is frequently referred to as an orchestration tool.

LangChain's growth has paralleled the explosive interest in generative AI, particularly following the launch of OpenAI's ChatGPT in late November 2022. ChatGPT demonstrated the remarkable capabilities of generative AI, rapidly gaining over 100 million users within just a few months. As developers sought to harness the power of these large language models, LangChain emerged as a critical framework, offering a more efficient and structured way to build and scale LLM-based applications.

The framework's rise reflects the broader trend of increasing demand for tools that can simplify and enhance the development of generative AI applications, making it a pivotal resource for AI developers worldwide.

Currently, LangChain averages over 15 million monthly downloads. It also powers over 100,000 apps, and the open source project has more than 75 stars, with over 3,000 contributors. Some of its customers include The Home Depot, Instacart, and Moody's.[1]

Let's take a look at a case study. It's for Ally Financial, which is a digital-only bank in the United States and has over 11 million customers.[2]

The company operates Ally.ai, which is an AI platform that provides assistance to more than 700 customer care associates. The technology handles tasks like summarizing conversations.

However, because of the stringent regulations for the financial services industry, Ally needed a way to protect customer information. The company did this by leveraging LangChain to build a tool to masked PII (personally identifiable information). Ally had five engineers work on this project for two months.

The results were significant. The system saved 2 minutes and 30 seconds per call and up to 85% of call summaries required no additional edits.

The Components

One of the key strengths of LangChain lies in its modularity. The framework is composed of several components, each designed to handle a specific aspect of AI development. This modular approach allows developers to mix and match components according to their specific needs, making it possible to build everything from simple

---

[1] https://www.langchain.com/

[2] https://blog.langchain.dev/ally-financial-collaborates-with-langchain-to-deliver-critical-coding-module-to-mask-personal-identifying-information-in-a-compliant-and-safe-manner/

chatbots to sophisticated AI agents capable of complex decision-making. Whether you're looking to create a model that can engage in multi-turn conversations, retrieve information from external sources, or execute a sequence of tasks autonomously, LangChain provides the building blocks needed to bring your vision to life.

In the next few sections, we'll take a look at the various components.

Models

Chat models are a fundamental component of LangChain, designed specifically to handle conversations by taking chat messages as inputs and returning chat messages as outputs, rather than merely processing plain text. This makes them particularly suited for building interactive and dynamic AI-driven applications. LangChain offers robust integrations with a variety of model providers, including OpenAI, Cohere, and Hugging Face, ensuring that developers can seamlessly interact with different models through a standardized interface.

In addition to basic functionality, LangChain enhances the usability of chat models by supporting multiple operational modes such as synchronous, asynchronous, batching, and streaming. This flexibility allows developers to choose the best mode for their specific use case, whether it be real-time interaction or processing large volumes of data. Moreover, LangChain provides additional features like caching, which can improve performance and reduce latency in applications.

Let's take a look at how to use a chat model. First, we need some setup:

```
pip install langchain

import openai
import os
import getpass

os.environ["OPENAI_API_KEY"] = getpass.getpass()
```

This code snippet is designed to set up the necessary environment to use LangChain with OpenAI's API. The first command, pip install langchain, installs the LangChain library.

The next part of the code imports the necessary libraries, including openai for interacting with OpenAI's models, os for handling environment variables, and getpass for securely capturing sensitive information like API keys. The os.environ["OPENAI_API_KEY"] = getpass.getpass() prompts the user to input their OpenAI API key securely without displaying it on the screen. This key is then stored as an environment variable, allowing the code to authenticate requests to OpenAI's API without exposing the key in the code itself.

Then we have this code:

```
from langchain_openai import ChatOpenAI

chat_model = ChatOpenAI(model="gpt-4o")
output = chat_model.invoke("What is LangChain?")
print(output.content)
```

The line from langchain_openai import ChatOpenAI imports the ChatOpenAI class from the langchain_openai module, which is designed to facilitate interactions with OpenAI's language models in a chat-like format.

By instantiating ChatOpenAI with chat_model = ChatOpenAI (model="gpt-4o"), the code creates an instance of the chat model configured to use OpenAI's "gpt-4o" version of GPT-4. This model is then ready to handle chat-based queries. The invoke method is used to send a question, in this case, "What is LangChain?", to the model. The model processes the input and generates a response, which is then stored in the output variable. Finally, print(output.content) is used to display the model's response, showcasing the ability of the chat model to generate meaningful and contextually appropriate replies.

Prompt Templates

The ChatPromptTemplate in LangChain is a tool designed to streamline the process of generating dynamic prompts for LLMs. This template class allows developers to create flexible and reusable prompt structures, where placeholders can be defined and later filled with specific values at runtime. This is particularly useful in scenarios where the content of the prompt needs to change based on the context or input provided by the user.

Here's a code sample:

```
from langchain_core.prompts import ChatPromptTemplate

# Create a new prompt template
prompt = ChatPromptTemplate.from_template(
    """List 3 benefits of using this technology:
{technology}"""
)

# Format the prompt with a specific technology
formatted_prompt = prompt.format(technology="blockchain")

# Get the response from the LLM
response = llm(formatted_prompt)

# Print out the response
print(response.content)
```

The process starts by importing ChatPromptTemplate from langchain_core.prompts, which allows the user to define a template with placeholders, such as {technology}, that can be filled in later with specific values.

In the example, a prompt template is created with the sentence "List 3 benefits of using this technology: {technology}." The placeholder {technology} is then replaced with "blockchain" using the format method, resulting in a fully customized prompt. This formatted prompt is passed to LLM to generate a response. Finally, the response content, which contains the AI-generated text, is printed out using `print(response.content)`.

Output Parsers

Output parsers play an important role in converting the output generated by LLMs into more structured and usable formats. This functionality is especially important when the task at hand involves generating structured data, such as JSON, XML, or CSV files. LangChain offers a wide array of output parsers, each designed to handle different types of outputs and formats, ensuring that the data produced is both accurate and easily interpretable.

One notable feature of LangChain's output parsers is their support for streaming, which allows for real-time processing and formatting of data as it is generated by the LLM. Additionally, many of these parsers come with format instructions to ensure the output adheres to a specified schema, although there are exceptions, such as when the schema is defined outside the prompt. Some output parsers, particularly those that handle complex or potentially error-prone outputs, can even call back to the LLM to correct any misformatted data, further enhancing the reliability and precision of the generated output. This capability makes LangChain's output parsers highly versatile tools for developers working with LLMs to create structured, actionable data from AI-generated content.

We'll take a look at a code example of an output parser. This will be by using a custom data structure defined with Pydantic. That is, we can ensure that the LLM-generated content is organized into specific fields such as product name, rating, and review details. For this, we'll use `JsonOutputParser`.

First, we have some setup:

```
from langchain_core.output_parsers import JsonOutputParser
from langchain_core.prompts import PromptTemplate
from langchain_core.pydantic_v1 import BaseModel, Field
# Initialize the ChatOpenAI model
model = ChatOpenAI(temperature=0)
```

We import JsonOutputParser and the PromptTemplate. Then we have BaseModel and Field from Pydantic. We will use these to define a structured data model, which ensures that the output conforms to specific data types and formats, such as strings for product names and integers for ratings. After this, we create an instance of the OpenAI LLM.

Then we create this class:

```
# Define the desired data structure for the product review
class ProductReview(BaseModel):
    product_name: str = Field(description="The name of the
    product being reviewed")
    rating: int = Field(description="The rating given to the
    product, out of 5")
    review_text: str = Field(description="The text of the review")
    pros: str = Field(description="The positive aspects
    mentioned in the review")
    cons: str = Field(description="The negative aspects
    mentioned in the review")
```

This code snippet defines a ProductReview class using Pydantic's BaseModel. Each attribute of the class—such as product_name, rating, review_text, pros, and cons—is defined with a specific data type (e.g., "str" for strings and "int" for integers) and accompanied by a descriptive Field. These descriptions clarify the expected content for each field, ensuring that when a language model generates a review, the output adheres to this predefined format.

We initialize a `JsonOutputParser` with the custom data structure defined by the ProductReview class:

```
parser = JsonOutputParser(pydantic_object=ProductReview)
```

We create a `PromptTemplate`:

```
prompt = PromptTemplate(
    template="Provide a detailed product review based on the
    user's input.\n{format_instructions}\n{query}\n",
    input_variables=["query"],
    partial_variables={"format_instructions": parser.get_
    format_instructions()},
)
```

The template string includes placeholders for `{format_instructions}` and `{query}`, where the former is dynamically filled with instructions from the `JsonOutputParser` on how to format the output according to the `ProductReview` schema. The `input_variables` parameter specifies that the prompt will take a user-provided query as input. By combining the prompt with format instructions and the user's query, this setup ensures that the LLM produces an output that is both relevant and structured according to the predefined format, making it easier to parse and use in downstream tasks.

We define the query for the LLM to generate a product review:

```
review_query = "Review the latest smartphone model XYZ."
```

We chain the prompt, model, and parser together:

```
chain = prompt | model | parser
```

This uses LangChain Expression Language (LCEL), which developers use to construct and manage complex AI workflows. LCEL provides a declarative way to compose LangChain components, allowing for more intuitive and flexible application design. With LCEL, developers can easily chain together various LangChain components, creating sophisticated pipelines that can handle complex tasks while maintaining readability and modularity.

In our code example, the pipe operator (|) allows you to chain together different components—starting with a PromptTemplate that generates a prompt, passing it to a language model (ChatOpenAI), and then feeding the model's output to a JsonOutputParser for structured processing. The pipe operator in LCEL creates a seamless and intuitive flow of data from one component to the next, enhancing readability and reducing the need for complex boilerplate code.

We run this with the following:

```
parsed_review = chain.invoke({"query": review_query})
```

Here, chain.invoke() is called with a dictionary containing the input variable, in this case, {"query": review_query}.

Then we print out the response, which will show the JSON output:

```
print(parsed_review)
```

Document Loaders

Document loaders allow for importing data from various sources and converting them into a standardized format known as Documents. A Document is essentially a piece of text accompanied by metadata, such as the source, creation date, or other relevant details. This uniform format allows for consistent processing and analysis, regardless of the text's origin. Document loaders are versatile and can handle a range of data sources, including .txt files, web page content, or even YouTube video transcripts, making it easy to integrate diverse textual data into a LangChain workflow.

Each document loader is equipped with a load method, which facilitates the conversion of data from a specified source into documents that can be used within the workflow. Additionally, some loaders offer a "lazy load" feature, which loads data only when necessary, optimizing memory usage for larger datasets. This feature is particularly useful when dealing with extensive data sources that don't need to be fully loaded at once.

Below is an example of how to use a document loader to read the contents of a CSV file:

```
from langchain_community.document_loaders.csv_loader import
CSVLoader
from tabulate import tabulate
```

The CSVLoader class is designed to facilitate the loading and processing of CSV files. Then we import the tabulate library. You will also need to install it using pip install tabulate. This library is for formatting and displaying tabular data.

Then we load the CSV file and prepare the data:

```
loader = CSVLoader(file_path='books_output.csv', source_
column="Book Title")
data = loader.load()
table_data = []
headers = ["Book Title"]
```

By specifying the file_path parameter as books_output.csv and using the source_column argument as "Book Title," this code ensures that each document generated from the CSV file is associated with a unique identifier based on the book title.

Once the data is loaded, the code prepares it for display using the tabulate library. The table_data list is initialized to hold the formatted data, and headers is set to ["Book Title"] to define the column name for the table.

Then we display the information:

```
for document in data:
    table_data.append([document.metadata['source']])

print(tabulate(table_data, headers=headers, tablefmt="grid"))
```

The for loop iterates over each document in the data list, which was previously loaded from the CSV file using the CSVLoader. For each document, the loop extracts the Book Title from the metadata['source'] attribute and appends it to the table_data list.

After collecting all the book titles, the code uses the tabulate library to print the data as a neatly formatted table. The tabulate function is called with table_data, headers, and tablefmt="grid" as arguments. headers contains the column name ("Book Title"), and tablefmt="grid" specifies the table format style, which renders the table with a grid-like appearance.

Text Splitters

When working with long documents in LangChain, it's often necessary to break them down into smaller, more manageable pieces. This process, known as text splitting, is crucial because it ensures that the text fits within the context window of your model, enabling efficient processing and analysis. LangChain provides a variety of built-in tools that allow users to split, combine, filter, and manipulate documents to suit specific application needs.

Text splitting is more complex than it might initially seem. The goal is to maintain the semantic integrity of the text, keeping related pieces of information together. What qualifies as "semantically related" can vary depending on the type of text you're working with. For instance, in some cases, it might be important to keep entire sentences or paragraphs intact, while in other scenarios, a more granular approach might be necessary. LangChain offers several methods to achieve this, ensuring that the resulting chunks of text are both meaningful and useful for the task at hand.

At a high level, text splitters in LangChain work by first breaking down the text into smaller, semantically meaningful units, often at the sentence level. These units are then combined into larger chunks until they reach a predefined size, based on specific measurement criteria. Once a chunk reaches the desired size, it is treated as a separate piece of text, and a new chunk begins, sometimes with overlapping content to maintain context across chunks. This process can be customized along two main axes: how the text is initially split and how the size of each chunk is measured.

LangChain offers a wide range of text splitters, each designed to handle different types of text and splitting needs. For example, there are splitters that focus on specific characters or tokens, as well as those tailored to particular formats like HTML or Markdown. More advanced splitters, like the RecursiveCharacterTextSplitter or the AI21SemanticTextSplitter, aim to keep semantically related text together by recursively splitting or identifying distinct topics within the text. Tools like Chunkviz can help visualize how these splitters operate, making it easier to fine-tune the splitting process to meet your specific requirements.

Text splitting is just one of the many document transformation capabilities in LangChain, with other tools available for integrating third-party services and performing additional modifications on documents before they are processed by language models.

Let's see an example. The `HTMLHeaderTextSplitter` is a specialized tool designed to process HTML content by breaking it down at the element level, such as headers, while preserving important structural information. This splitter is particularly useful for documents where the structure is essential for maintaining context, such as web pages or structured reports. By splitting the text based on HTML headers, the splitter not only divides the content into manageable chunks but also adds metadata for each header, making it easier to keep related sections of text together and maintain their semantic meaning.

191

We have some setup code:

```
pip install lxml
from langchain_text_splitters import HTMLHeaderTextSplitter
```

To work with HTML content in Python, you may need to install the "lxml" library, which is a powerful and efficient tool for parsing and manipulating HTML and XML documents. Once installed, you can use it in conjunction with tools like the `HTMLHeaderTextSplitter` from the `langchain_text_splitters` module. This splitter allows you to process HTML content by breaking it into segments based on specific header tags (like "<h1>," "<h2>," etc.), making it easier to analyze and manipulate structured text within an HTML document.

Then we read the HTML file:

```
with open('sample_document.html', 'r', encoding='utf-8')
as file:
    html_string = file.read()
```

This code opens the file sample_document.html in read mode ("r") with UTF-8 encoding, ensuring that any special characters are correctly interpreted. The `with open(...)` construct is used for opening the file, which ensures that the file is properly closed after reading, even if an error occurs. The `file.read()` method reads the entire content of the file into the variable `html_string`, which can then be used for further processing, such as parsing or splitting the HTML content.

Next, we define the headers to split on:

```
headers_to_split_on = [
    ("h1", "Header 1"),
    ("h2", "Header 2"),
    ("h3", "Header 3"),
]
```

Then we work with the data:

```
html_splitter = HTMLHeaderTextSplitter(headers_to_split_
on=headers_to_split_on)
html_header_splits = html_splitter.split_text(html_string)
print(html_header_splits)
```

First, an instance of HTMLHeaderTextSplitter is created and configured with the headers defined in the headers_to_split_on list. This list specifies which HTML tags (such as "<h1>," "<h2>," "<h3>") should be used as breakpoints for splitting the text. Then, the split_text method is called on the html_splitter object, passing in the html_string containing the HTML content. This method processes the HTML content and divides it into segments wherever the specified headers appear. The resulting segments are stored in html_header_splits, which is then printed to display the split sections of the HTML document. This approach is useful for extracting and organizing structured content from HTML files, making it easier to analyze or further manipulate the data.

Memory

In most LLM applications, a conversational interface is key, allowing the system to engage in meaningful dialogue with users. A vital aspect of any conversation is the ability to reference information that was shared earlier in the discussion. At the very least, a conversational system should be capable of accessing a portion of past messages. For more advanced systems, it is essential to have a continually updated world model that tracks entities and their relationships throughout the conversation.

This capability to retain and utilize information from previous interactions is known as memory. LangChain offers numerous tools to incorporate memory into a system. These utilities can function independently or be integrated smoothly into a chain of processes. However, much of the memory-related functionality in LangChain is currently in beta, as it is still under development and may not yet be suitable for production environments. Most of this functionality also operates with legacy chains rather than LCEL syntax.

A notable exception is the ChatMessageHistory feature, which is largely production-ready and compatible with LCEL. ChatMessageHistory allows the system to store and retrieve past messages, thereby maintaining context across interactions. This feature supports both reading and writing actions, meaning a system can augment user inputs based on previous interactions and store new interactions for future reference.

When building a memory system, two primary design decisions must be made: how the state (or memory) is stored and how it is queried. At the core of any memory system is a history of chat interactions. These interactions need to be stored, whether in-memory or in persistent storage like databases. LangChain's memory module offers a range of integrations for managing chat message storage, from simple in-memory lists to more robust database solutions.

Once messages are stored, the challenge is how to query this stored information effectively. A basic memory system might return the most recent messages, while a more complex system could summarize past interactions or extract specific entities mentioned in previous conversations. The LangChain memory module aims to be flexible, allowing developers to start with simple memory systems and customize them as needed for specific applications.

Let's look at a code example that uses memory. This will be a chatbot designed to remember details about the user's career goals and previous discussions, helping to provide more personalized advice in future interactions.

We will start with preliminary setup:

```
from langchain_core.messages import SystemMessage
from langchain_core.prompts import (
    ChatPromptTemplate,
    HumanMessagePromptTemplate,
    MessagesPlaceholder,
)
from langchain_openai import ChatOpenAI
from langchain.memory import ConversationBufferMemory
from langchain.chains import LLMChain
```

We import a variety of LangChain Templates like SystemMessage, ChatPromptTemplate, and HumanMessagePromptTemplate. These help to structure the conversational prompts by setting a system-wide context and defining how user inputs are incorporated. MessagesPlaceholder is used to insert conversation history into the prompt, allowing for continuity across interactions. ChatOpenAI represents the chat-based language model used to generate responses, while ConversationBufferMemory manages the storage of previous chat messages, enabling the model to reference past interactions. Finally, LLMChain ties these components together, creating a functional chain that processes user inputs, maintains context through memory, and generates coherent responses in a conversation.

We then set up the chat prompt:

```
prompt = ChatPromptTemplate.from_messages(
    [
        SystemMessage(
            content="You are a career coach chatbot, helping
            users with their career goals."
        ),  # The persistent system prompt
        MessagesPlaceholder(
            variable_name="chat_history"
```

195

```
    ),  # Where the memory will be stored
    HumanMessagePromptTemplate.from_template(
        "{human_input}"
    ),  # Where the human input will be injected
  ]
)
```

This code snippet constructs a `ChatPromptTemplate` for a chatbot designed to act as a career coach. The `SystemMessage` sets a consistent and persistent context, indicating that the chatbot's role is to help users with their career goals. The `MessagesPlaceholder` is used to insert the chat history, allowing the chatbot to remember and reference previous interactions during the conversation. Finally, the `HumanMessagePromptTemplate` is configured to dynamically inject the user's current input into the prompt, ensuring that the chatbot can respond appropriately to each new message while maintaining the overall context of the conversation. This setup enables the chatbot to deliver personalized and context-aware career advice.

Next, we initialize the memory to keep track of the conversation history and then initialize the OpenAI mode:

```
memory = ConversationBufferMemory(memory_key="chat_history",
return_messages=True)

llm = ChatOpenAI()
```

We then create the LLMChain with the prompt, memory, and model:

```
chat_llm_chain = LLMChain(
    llm=llm,
    prompt=prompt,
    verbose=True,
    memory=memory,
)
```

We then simulate a conversation, to see how the memory works with the chatbot:

```
response = chat_llm_chain.predict(human_input="I'm thinking
about switching careers to data science.")
print(response)

response = chat_llm_chain.predict(human_input="What skills do I
need to develop for a data science role?")
print(response)

response = chat_llm_chain.predict(human_input="I have started
learning Python. What should I focus on next?")

print(response)
```

The first line sends the user's input, "I'm thinking about switching careers to data science," to the chatbot, and the response is stored in the response variable and printed out. This initiates the conversation and sets the context for future interactions. The second line continues the conversation by asking, "What skills do I need to develop for a data science role?" Again, the chatbot generates a response based on both the current input and the previous context, which is also printed. Finally, the third line further develops the discussion by stating, "I have started learning Python. What should I focus on next?" The chatbot, remembering the prior exchanges, provides advice tailored to the user's ongoing journey into data science, and the response is printed. Each of these steps demonstrates how the chatbot uses memory to deliver increasingly personalized and context-aware career advice.

Key Concepts of LangChain Agents

Creating AI agents in LangChain involves several key concepts and components that allow for dynamic decision-making and interaction within the AI system. Agents in LangChain use language models to determine the sequence of actions dynamically.

Here's a look at the core concepts:

- AgentAction: This is a critical component that defines the action an AI agent should take at a particular step. Each AgentAction consists of a tool that the agent should invoke and the input that should be passed to this tool. This allows the agent to interact with various tools, such as APIs or databases, and retrieve or manipulate data as needed.

- AgentFinish: When an agent has completed its task and is ready to return the result, it uses the AgentFinish component. This contains the final output of the agent, typically structured as a key-value mapping. The output is usually a string that represents the agent's final response to the user, encapsulating all the steps taken during the interaction.

- Intermediate Steps: These represent the history of actions and corresponding outputs during the current run of the agent. By maintaining this history, the agent can track what has already been done, which is essential for ensuring that future steps are relevant and do not duplicate previous work. This is represented as a list of tuples, where each tuple contains an AgentAction and its associated output.

- Agent: The agent itself is responsible for determining the next action to take based on the current context and the intermediate steps taken so far. It typically relies on a language model to make these decisions, using a prompt to encode the inputs and an output parser to interpret the results. Agents can be customized with different prompting styles, input encodings, and output parsing strategies to suit specific tasks.

- AgentExecutor: The agent executor is the runtime environment where the agent operates. It continuously cycles through the process of selecting actions, executing them, and processing their outputs until the agent reaches a conclusion. The executor handles complexities such as error management, tool selection, and logging, ensuring the smooth operation of the agent.

Types of Agents

LangChain offers a variety of agent types, each designed to handle specific tasks and workflows within AI systems. These agents leverage different strategies for decision-making and interaction, enabling developers to tailor their AI solutions to various use cases.

They include the Tool Calling Agent, XML Agent, JSON Chat Agent, Structured Chat Agent, ReAct Agent, and Self-Ask with Search Agent.

Tool Calling Agent

The Tool Calling Agent is a versatile agent that allows an LLM to determine when and how to invoke external tools. By defining a set of tools that the agent can use, the model can intelligently decide which tool to call based on the inputs it receives. The agent then produces structured outputs, such as JSON objects, that specify the arguments needed to call these tools. This approach enhances the reliability and accuracy of tool calls compared to using a generic text completion or chat API. The Tool Calling Agent supports a wide range of providers, including OpenAI, Anthropic, Google Gemini, and Mistral, making it a powerful and flexible solution for integrating tool usage into AI workflows.

XML Agent

The XML Agent is tailored for LLMs, such as Anthropic's Claude, that excel at reasoning and writing in XML format. This agent is particularly useful when working with unstructured tools that accept a single string input. By structuring its outputs in XML, the agent can interact more effectively with certain models and tools, providing a specialized solution for tasks that require XML-based reasoning.

JSON Chat Agent

The JSON Chat Agent is designed for language models that are proficient at generating JSON-formatted outputs. This agent is particularly useful when working with chat models that require structured data exchanges. By using JSON to format its outputs, the agent can support complex interactions with chat models, making it easier to manage and process multi-step conversations or data-intensive tasks.

Structured Chat Agent

The Structured Chat Agent is capable of handling multi-input tools, making it ideal for scenarios where multiple pieces of information need to be processed simultaneously. This agent type supports more complex interactions, allowing it to manage tasks that require the integration of various data points or inputs in a structured manner. It is particularly useful in applications where the agent must coordinate multiple actions or decisions based on a combination of inputs.

Self-Ask with Search Agent

The Self-Ask with Search Agent is designed for tasks that require iterative querying and information retrieval. This agent uses a process of self-questioning, where it asks itself clarifying questions to better understand the task before conducting a search for the necessary information. This approach is particularly useful for tasks that involve complex information retrieval or where the agent needs to gather data from external sources to complete a query.

Next, we will take a deeper look at the ReAct Agent and see how to code one using LangChain.

ReAct Agent

In October 2022, researchers from Google Research and Princeton University published a paper entitled, "ReAct: Synergizing Reasoning and Acting in Language Models."[3] This established a new type of agent called ReAct, which has become critical for LangChain. It's about helping LLMs to reason and act to solve general tasks.

Before ReAct, the abilities of LLMs to these activities were mostly studied separately. But this led to disappointing results. ReAct's combination or synergizing of reasoning and tasks has meant significant improvements with agents.

---

[3] https://arxiv.org/abs/2210.03629

In the paper, the researchers used Google's PaLM-540B and fed it both specific actions (like "search" or "go to") and reasoning traces (logical steps or thoughts) for solving tasks. Depending on the task, the model alternated between generating reasoning traces and actions, creating a road map or trajectory to solve the task. The ReAct dynamically adjusted its plans based on new information and interacted with external sources like Wikipedia for more data. There are various types of reasoning traces employed like creating action plans, injecting commonsense knowledge, and adjusting plans when faced with exceptions. This way, the model wasn't just thinking about the task but also taking steps to solve it.

Agent Program

This program is a project management assistant that integrates tools for task status retrieval and online documentation search. It utilizes a structured approach where specific tools are defined to either look up the status of tasks in an internal database or search for relevant documentation online using DuckDuckGo. The program leverages an OpenAI LLM to guide the interaction, with the input being processed according to a custom prompt that directs the agent's responses. The agent is capable of performing actions like retrieving the status of a task based on its ID and searching for best practices or other project-related documentation, making it a useful tool for project managers or team members needing quick access to task-related information.

We start with these imports:

```
from langchain import PromptTemplate
from langchain.tools import StructuredTool
from langchain_openai import ChatOpenAI
from langchain.agents import create_tool_calling_agent,
AgentExecutor
```

```
from langchain.tools import DuckDuckGoSearchResults
from pydantic import BaseModel
```

PromptTemplate is used to create customizable prompts that guide the language model's behavior. The StructuredTool is a class that allows you to define tools that can handle structured input data, making it easier to interact with specific functions or APIs. ChatOpenAI is an interface to OpenAI's language models, enabling natural language processing capabilities within the program. The create_tool_calling_agent and AgentExecutor from langchain.agents are used to create and manage an agent that can decide when and how to use the defined tools based on user input. DuckDuckGoSearchResults provides a tool for searching the Web via DuckDuckGo, which can be integrated into the agent for tasks like finding online documentation. Finally, BaseModel from pydantic is used to define data models that enforce type validation and structured input, ensuring that the tools receive correctly formatted data.

After this, we define the task status retriever tool:

```
class TaskStatusRetriever:
    def __init__(self, task_database):
        self.task_database = task_database

    def get_status(self, task_id: str) -> str:
        return self.task_database.get(task_id, "Task
not found")

task_database = {
    "task1": "In Progress",
    "task2": "Completed",
    "task3": "Not Started",
}
```

```
task_status_retriever = TaskStatusRetriever(task_database)

def task_status_lookup(task_id: str) -> str:
    return task_status_retriever.get_status(task_id)
```

This code defines a `TaskStatusRetriever` class that simulates a system for retrieving the status of tasks from a project management database. The class is initialized with a `task_database`, a dictionary that stores task IDs as keys and their statuses as values. The `get_status` method is used to look up the status of a specific task by its ID, returning the status if found or "Task not found" if the task ID does not exist in the database. The `task_database` dictionary contains example tasks with predefined statuses ("In Progress," "Completed," and "Not Started"). The `task_status_retriever` object is an instance of this class, and the `task_status_lookup` function serves as an interface to retrieve the status of a task by passing in its ID, making it easy to integrate this functionality into a larger system.

Next, we'll define the input schema for TaskStatusLookup:

```
class TaskStatusInput(BaseModel):
    task_id: str

task_status_tool = StructuredTool.from_function(
    name="TaskStatusLookup",
    description="Look up the status of a task by its ID.",
    func=task_status_lookup,
    args_schema=TaskStatusInput
)
```

This code defines a `TaskStatusInput` class using Pydantic's `BaseModel`, which serves as a schema to enforce the structure and type of input data for the task status lookup tool. The `TaskStatusInput` model specifies that the input must include a `task_id` as a string. The `task_status_tool` is then created using `StructuredTool.from_function`, which generates a tool based on the `task_status_lookup` function. This tool is named

"TaskStatusLookup" and is described as a tool to look up the status of a task by its ID. By using the TaskStatusInput model as the args_schema, the tool ensures that it receives properly formatted input, thereby facilitating structured and validated interaction with the task status retrieval functionality within a larger LangChain-based application.

Then we'll use the Online Documentation Search Tool:

```
duckduckgo_search = DuckDuckGoSearchResults()

def search_docs(query: str) -> str:
    return duckduckgo_search.run(query)
```

This code initializes a DuckDuckGoSearchResults object, which provides the ability to perform web searches using the DuckDuckGo search engine within a LangChain application. The search_docs function is defined to take a search query as input, represented by the query string. When called, the function uses the duckduckgo_search object to execute the search by invoking its run method with the provided query. The function returns the search results as a string.

We define the input schema for DocsSearch:

```
class DocsSearchInput(BaseModel):
    query: str

docs_search_tool = StructuredTool.from_function(
    name="DocsSearch",
    description="Search for relevant documentation online.",
    func=search_docs,
    args_schema=DocsSearchInput
)
```

The DocsSearchInput model includes a single field, query, which is a string representing the search query. The docs_search_tool is then created using StructuredTool.from_function, which constructs a tool based on the search_docs function. This tool is named "DocsSearch" and is described as a tool to search for relevant documentation online. By using the DocsSearchInput model as the args_schema, the tool ensures that the input is validated and correctly structured.

We create the Tools List:

```
tools = [task_status_tool, docs_search_tool
```

Then we define the LLM and the prompt:

```
llm = ChatOpenAI(model="gpt-4o-mini", temperature=0)

prompt_template = """
You are a project assistant. You can look up the status of
tasks in a project management system and search for relevant
documentation online.
Respond based on the user's input using the appropriate tools.

User's input: {input}

{agent_scratchpad}
"""

prompt = PromptTemplate.from_template(prompt_template)
```

The llm variable is initialized with an instance of ChatOpenAI, specifying the use of the "gpt-4o-mini" model from OpenAI. The temperature=0 parameter is set to make the model's outputs more deterministic, reducing randomness in its responses.

The prompt_template is a string that defines the structure and instructions for how the agent should behave. It describes the agent as a project assistant capable of looking up task statuses in a project management system and searching for relevant documentation online. The template includes placeholders {input} for the user's input and {agent_scratchpad} for the agent's internal reasoning or intermediate steps.

Finally, the prompt variable is created by passing the prompt_template string to PromptTemplate.from_template, which formats the prompt for use with the language model. This setup ensures that the agent responds appropriately to user queries by leveraging the language model and the defined tools.

Then we create the agent as well as the Agent Executor. Then we test it and create the response:

```
agent = create_tool_calling_agent(llm, tools, prompt)

agent_executor = AgentExecutor(agent=agent, tools=tools,
verbose=True)

response = agent_executor.invoke({"input": "What's the status
of task1?"})
print(response['output'])

response = agent_executor.invoke({"input": "Find documentation
on project management best practices"})
print(response['output'])
```

First, the agent is created using the create_tool_calling_agent function, which combines the language model (llm), the list of tools (tools), and the prompt template (prompt). This agent is designed to interpret user input, determine the appropriate tool to use, and provide a response based on the results from that tool.

Next, the agent_executor is initialized using the AgentExecutor class. The agent_executor manages the execution of the agent, handling the sequence of actions the agent takes when responding to user queries. The verbose=True argument ensures that the execution process is logged, providing detailed output for debugging or understanding the agent's behavior.

The agent is then tested with two example queries. The `invoke` method is called with a dictionary containing the user's input under the "input" key. The first query asks for the status of a specific task, task1, and the agent's response is printed. The second query asks for documentation on project management best practices, and, again, the agent's response is printed.

The output from the agent demonstrates its ability to process user queries by utilizing the appropriate tools. In the first instance, the agent successfully retrieves the status of a specific task (`task1`), confirming that it is "In Progress." In the second instance, the agent searches for documentation on project management best practices and provides a list of relevant online resources. These include links and brief descriptions of articles from reputable sources, which cover various project management techniques and methodologies. The output shows that the agent effectively switches between tools to deliver accurate and helpful responses based on the user's input.

Conclusion

LangChain's modularity and support for both Python and JavaScript make it a foundational tool for AI developers. The chapter highlighted the importance of LangChain's orchestration abilities, its rapid growth following the rise of generative AI, and its adoption by major companies. The chapter also dived into the components of LangChain, including models, prompt templates, output parsers, document loaders, and text splitters, providing practical examples and code snippets to illustrate their use. The chapter then concluded with LangChain agents, explaining how to build them to handle dynamic decision-making and interaction within AI systems.

CHAPTER 9

Introduction to LangGraph

LangGraph is a popular open source framework—created by LangChain—that helps developers use large language models (LLMs) to build sophisticated, stateful, and multi-actor applications. This capability is vital for crafting advanced agent architectures capable of context retention, learning from interactions, and continuous evolution.

A key distinction of LangGraph lies in its departure from the limitations of Directed Acyclic Graph (DAG) structures, which are common in many LLM frameworks. They are based on a conceptual model used in computer science and mathematics. It consists of a finite set of vertices (or nodes) connected by directed edges, with the crucial constraint that there are no cycles or loops within the structure. In other words, if you start at any node and follow the directed edges, you can never return to the same node. This structure is widely used in task scheduling, data processing pipelines, and dependency resolution, where a clear, one-way flow of operations or data is required.

While DAGs are useful for many applications, they have limitations when it comes to creating truly dynamic and adaptive AI agents. This is where LangGraph's innovative approach comes into play. By enabling the creation of cycles within the workflow, LangGraph opens up possibilities for implementing iterative processes, feedback loops, and recursive behaviors—all essential components of genuine agentic intelligence.

© Tom Taulli, Gaurav Deshmukh 2025
T. Taulli and G. Deshmukh, *Building Generative AI Agents*,
https://doi.org/10.1007/979-8-8688-1134-0_9

LangGraph's architectural inspiration comes from established frameworks like Pregel and Apache Beam, with its public interface drawing from NetworkX concepts. This fusion results in a powerful yet accessible AI development tool. While LangGraph is designed to work seamlessly with LangChain and LangSmith, it maintains the flexibility to be used independently, catering to developers with diverse toolchain preferences.

Regardless, it is still important to understand the fundamentals of LangChain. This will be the focus of the first half of this chapter. After this, we'll dive into how LangChain works.

Benefits of Combining LangChain with LangGraph

The integration of LangChain and LangGraph represents a significant advancement in the field of AI development, creating a powerful ecosystem that offers developers unprecedented capabilities. This synergy combines LangChain's extensive toolset for large language model (LLM) interactions with LangGraph's sophisticated stateful framework, resulting in a comprehensive solution for building advanced AI applications.

At the core of this integration is the ability to handle complexity with greater ease and efficiency. Developers can now create AI agents capable of managing intricate, multi-step processes that were previously challenging to implement. These agents benefit from improved context awareness, maintaining state across interactions, learning from experiences, and evolving over time. This dynamic adaptability leads to more intelligent and responsive AI systems that can handle a wide range of tasks with increased sophistication.

The combined framework offers a streamlined development experience, particularly beneficial for those already familiar with LangChain. The shared concepts and patterns between the two systems reduce the learning curve, allowing developers to quickly leverage LangGraph's advanced features. This continuity accelerates the development process, enabling faster creation and deployment of complex AI agents. Furthermore, the integration provides developers with increased flexibility, offering a broader palette of tools and approaches to choose from when solving specific AI challenges.

One of the key strengths of this integration is the enhanced workflow management capabilities. LangGraph's cyclic workflows complement LangChain's robust language model utilities, enabling the implementation of sophisticated feedback loops and iterative behaviors within AI agents. This feature is crucial for creating AI systems that can refine their responses and adapt their strategies based on ongoing interactions and outcomes.

The scalability offered by this combined approach is another significant advantage. As AI applications grow in complexity and data volume, the integrated framework provides the necessary tools and structures to scale effectively. This scalability is complemented by improved error handling mechanisms, enhancing the overall reliability and robustness of AI systems. The modular nature of the combined framework also facilitates easier maintenance and updates, allowing developers to create more reusable components.

From a resource perspective, the LangChain-LangGraph integration enables more efficient use of computational resources. This optimization can lead to reduced costs in AI application deployment, making advanced AI solutions more accessible to a broader range of projects and organizations. The framework also expands the potential use cases, enabling developers to tackle a wider spectrum of AI challenges, from straightforward chatbots to complex decision-making systems that require nuanced understanding and reasoning.

Testing and debugging complex AI systems become more manageable with this integrated approach. The combined framework offers enhanced tools for examining AI agent behaviors and troubleshooting intricate workflows. This improvement in the development and quality assurance process leads to more reliable and performant AI applications.

Lastly, the integration enhances interoperability, making it easier to connect AI agents with external systems and data sources. This expanded connectivity opens up new possibilities for AI applications, allowing them to integrate more seamlessly into existing technological ecosystems and leverage a wider range of data and functionalities.

In conclusion, the combination of LangChain and LangGraph provides developers with a comprehensive toolkit that pushes the boundaries of what's possible in AI development. It enables the creation of more sophisticated, adaptable, and context-aware AI applications, setting a new standard for intelligent systems. As this integrated approach continues to evolve, it promises to drive innovation in AI, opening up new frontiers in machine intelligence and its practical applications across various domains.

Pros and Cons of LangGraph

One of the standout features of LangGraph is its emphasis on controllability. As a low-level framework, it offers developers fine-grained control over both the flow of operations and the state of the application. This level of control is crucial for creating reliable and predictable AI agents, especially in scenarios where precision and consistency are paramount. Developers can define complex decision trees, implement conditional logic, and orchestrate multi-step processes with a high degree of specificity.

Persistence is another key advantage of LangGraph. The framework includes built-in mechanisms for automatically saving the state after each step in the graph. This feature opens up a world of possibilities for advanced applications. It enables seamless implementation of human-in-the-loop workflows, where human operators can intervene, provide feedback, or make decisions at critical junctures. The persistence capability also facilitates error recovery, allowing developers to pause and resume graph execution at any point. This is particularly valuable for long-running processes or in scenarios where reliability and fault tolerance are critical.

LangGraph's design philosophy draws inspiration from established frameworks like Pregel and Apache Beam, while its public interface borrows concepts from NetworkX. This blend of influences results in a powerful yet accessible tool for AI development. While LangGraph is built to integrate seamlessly with LangChain and LangSmith, it's important to note that it can be used independently, offering flexibility to developers who may have different toolchain preferences.

The framework also shines in its support for streaming outputs. As each node in the graph produces results, these can be streamed in real time, including token-by-token streaming from language models. This feature is invaluable for creating responsive, interactive AI agents that can provide immediate feedback and engage in dynamic conversations.

In the broader context of AI development, LangGraph represents a significant step forward in the creation of more sophisticated, stateful AI agents. By providing tools for implementing cycles, ensuring persistence, and offering fine-grained control, it empowers developers to create AI systems that can handle complex, multi-step tasks, maintain context over extended interactions, and adapt to changing conditions. This makes it particularly well suited for applications in areas such as conversational AI, task planning and execution, and multi-agent simulations.

As AI continues to evolve and find new applications across industries, frameworks like LangGraph play a crucial role in bridging the gap between the raw capabilities of large language models and the complex, real-world requirements of AI systems. By providing a robust foundation for building stateful, adaptive AI agents, LangGraph is poised to accelerate innovation in the field of artificial intelligence and enable the creation of more capable, reliable, and sophisticated AI applications.

Graphs

LangGraph models agent workflows as graphs, where the behavior of your agents is defined by three essential components. First, the State represents the current snapshot of your application, which can be any Python type, though it often takes the form of a TypedDict or Pydantic BaseModel. Next, Nodes are Python functions that encode the logic of your agents, taking the current State as input, performing some computation or action, and returning an updated State. Lastly, Edges are Python functions that determine the next Node to execute based on the current State, guiding the flow of operations through either conditional branches or fixed transitions.

By combining Nodes and Edges, you can create intricate, looping workflows that allow the State to evolve over time. The real strength of LangGraph lies in how it manages this State, with Nodes and Edges functioning as Python code—whether incorporating an LLM or utilizing standard Python logic.

In essence, Nodes perform the tasks, while Edges dictate the next steps. LangGraph's underlying graph algorithm employs message passing to define a general program structure. When a Node completes its operation, it sends messages along its Edges to subsequent Nodes, which then execute their functions and pass the resulting messages onward. This process continues in a pattern inspired by Google's Pregel system, advancing through discrete "super-steps." Each super-step corresponds to a single iteration over the graph's nodes, where parallel operations belong to the same super-step, while sequential ones are divided across separate super-steps.

At the start of graph execution, all nodes are inactive, becoming active only when they receive new messages (state) through their incoming edges or channels. An active node runs its function and sends updates in response. Once a super-step concludes, nodes without incoming messages signal their inactivity. The graph's execution completes when all nodes are inactive, and no messages remain in transit.

The StateGraph class is the primary graph class used, parameterized by a user-defined State object. On the other hand, the MessageGraph class is a specialized type of graph where the State is merely a list of messages, making it rarely used except in chatbot applications, where the State's complexity is minimal.

To construct your graph, you first define the State, then add nodes and edges, and finally compile it. Compiling is a straightforward process that performs basic structural checks on your graph, ensuring there are no orphaned nodes, among other things. It also allows you to specify runtime arguments such as checkpointers and breakpoints. Compiling is done by calling the ".compile" method on your graph.

In the next few sections, we'll take a deeper look at State, Nodes, and Edges.

State

The State includes the schema and reducer functions that dictate how updates are applied. The schema, which serves as the input for all Nodes and Edges, can be a TypedDict or a Pydantic model. Nodes emit updates to the State, which are then processed using the specified reducer functions.

The schema is typically defined using TypedDict, but a Pydantic BaseModel can also be used to incorporate default values and additional data validation. By default, the input and output schemas of the graph are the same, but you can customize them if needed, particularly when handling numerous keys with distinct roles.

Reducers play a crucial role in applying updates to the State. Each key in the State has an independent reducer function, with the default behavior being to overwrite the key with new updates. For instance, you can use the Annotated type to specify a custom reducer, like operator.add, which can append updates rather than overwrite them.

Context channels allow you to manage shared resources like database connections that are maintained outside the nodes and excluded from checkpointing. These resources are set up at the beginning of the graph execution and cleaned up at the end, ensuring efficient management throughout the graph's run.

When working with messages in your graph's State, particularly in applications involving chat models, it's beneficial to store conversation history as a list of Message objects. By adding a key to the State for these messages and using a reducer like operator.add, you can efficiently manage message updates. Alternatively, the add_messages function can be used to track message IDs and handle both new and updated messages accurately, ensuring that manual updates don't inadvertently append messages but instead update them as needed.

For serialization, add_messages also facilitates the deserialization of messages into LangChain Message objects, allowing seamless state updates. You can easily access these messages using dot notation, like state["messages"][-1].content.

Nodes

Nodes are typically Python functions, either synchronous or asynchronous, where the first argument is the state and the optional second argument is a "config" that holds configurable parameters, such as a session_id. You add these nodes to a graph using the add_node method.

For example, consider a function process_data that logs a message and returns a modified state:

```
from langchain_core.runnables import RunnableConfig
from langgraph.graph import StateGraph

builder = StateGraph(dict)
```

```
def process_data(state: dict, config: RunnableConfig):
    print("Processing data for session:",
    config["configurable"]["session_id"])
    state["output"] = f"Processed: {state['input']}"
    return state

def finalize_data(state: dict):
    state["status"] = "complete"
    return state

builder.add_node("process_data", process_data)
builder.add_node("finalize_data", finalize_data)
```

When these functions are added to the graph, they are automatically converted into RunnableLambda objects, which provide features like batch processing, asynchronous execution, and built-in tracing and debugging. If you add a node without specifying a name, the function's name is used as the node name by default.

Special nodes such as START and END are crucial for controlling the flow within the graph. The START node is used to designate the entry point for user input, allowing you to specify which nodes are triggered first:

```
from langgraph.graph import START
builder.add_edge(START, "process_data")
```

The END node serves as a terminal point in the graph, marking where no further actions will occur after the execution of its associated edges:

```
from langgraph.graph import END
builder.add_edge("finalize_data", END)
```

These special nodes help define the structure of your graph, ensuring clear entry and exit points for the data flow.

Edges

There are several key types of edges:

- Normal Edges: Directly connect one node to the next

- Conditional Edges: Use a function to decide which node(s) to transition to next

- Entry Point: Specifies the first node to execute when user input is received

- Conditional Entry Point: Uses a function to determine the initial node(s) to execute based on custom logic

A node can have multiple outgoing edges, meaning all destination nodes will execute in parallel in the next super-step.

For a normal edge, you will use the add_edge method:

```
graph.add_edge("step_one", "step_two")
```

For more complex routing, you can use a conditional edge. This uses a method that requires a node name and a routing function, which uses the current state of the graph to determine the next node or nodes:

```
def routing_logic(state):
    return "step_two" if state["condition"] else "step_three"

graph.add_conditional_edges("step_one", routing_logic)
```

You can also map the function's output to specific nodes:

```
graph.add_conditional_edges("step_one", routing_logic, {True:
"step_two", False: "step_three"})
```

For an Entry Point Edge, you connect the virtual START node to the first node using add_edge:

```
from langgraph.graph import START

graph.add_edge(START, "initial_step")
```

Finally, with a Conditional Entry Point, you need to start at different nodes based on some condition. For this, you can use add_conditional_edges from the START node:

```
from langgraph.graph import START

def start_logic(state):
    return "initial_step" if state["start_here"] else
    "alternative_step"

graph.add_conditional_edges(START, start_logic, {True:
"initial_step", False: "alternative_step"})
```

This approach allows you to dynamically control the flow based on the incoming data, ensuring that the graph starts in the most appropriate place, depending on the situation.

Reflection Agent

The reflection agent in LangGraph is a specialized type of agent designed to analyze and evaluate its own decisions and actions, enabling it to improve performance over time. Unlike reactive agents that respond to inputs in a straightforward manner, reflection agents incorporate a layer of self-assessment, allowing them to learn from past interactions and outcomes. This self-reflection capability is crucial for tasks that require ongoing optimization, such as content creation or strategy development. By iterating on its decisions and considering what worked well and what didn't, a reflection agent can refine its output to achieve higher-quality results over time.

To demonstrate this, we'll create a program that makes better tweets.

We will first need to install various libraries, including those for OpenAI, LangChain, and LangGraph:

```
pip install --upgrade -q openai langchain langchain-openai
langchain-community langgraph
```

Next, we will have these imports:

```
from typing import TypedDict, Annotated, Sequence
from langgraph.graph import Graph, StateGraph
from langchain_openai import ChatOpenAI
from langchain_core.prompts import PromptTemplate
from langchain_core.output_parsers import StrOutputParser
```

The import statement `from typing import TypedDict, Annotated, Sequence` brings in important type hinting features from Python's `typing` module. `TypedDict` is used to create dictionaries with a predetermined structure, where each key is associated with a specific type, ensuring that data follows a consistent format. `Annotated` allows you to add metadata or constraints to types, which can be useful for enhancing type hints and making them more informative for tools or frameworks. `Sequence` is a type hint that represents ordered collections, such as lists or tuples, allowing you to specify the type of elements contained within the sequence.

From the LangGraph library, the imports from `langgraph.graph import Graph, StateGraph` introduce two essential components, `Graph` and `StateGraph`.

Then there are various libraries from LangChain, such as `ChatOpenAI` for integrating OpenAI's chat models, `PromptTemplate` for managing structured prompts, and `StrOutputParser` for parsing string outputs.

We'll then initialize the LLM:

```
llm = ChatOpenAI(model="gpt-4o-mini")
```

We will define the prompt templates:

```
reflection_template = PromptTemplate.from_template(
    """
    Analyze the following tweet and provide a reflection on how
    it can be improved:
    Tweet: {tweet}

    Consider aspects such as clarity, engagement, and brevity.
    Provide specific suggestions.
    """
)
```

The template, named reflection_template, is designed to analyze tweets and provide reflections on how they can be improved. It asks for an analysis of a given tweet, focusing on aspects like clarity, engagement, and brevity. By using the PromptTemplate.from_template method, the template is created with placeholders, such as {tweet}, which can be dynamically filled with specific tweet content during runtime. This approach ensures that the prompt is consistently formatted, making it easier to generate targeted and effective reflections for tweet improvement.

This function is to improve the tweet:

```
improve_tweet_template = PromptTemplate.from_template(
    """
    Given the original tweet and the reflection, provide an
    improved version of the tweet:

    Original tweet: {tweet}
    Reflection: {reflection}

    Improved tweet:
    """
)
```

The placeholders {tweet} and {reflection} allow for dynamic insertion of the original tweet content and the analysis provided by the reflection, respectively. This template ensures that the process of refining tweets is consistent and guided by specific feedback, making it easier to produce more effective and engaging tweets.

We define the nodes:

```
def reflect(state: AgentState) -> AgentState:
    chain = reflection_template | llm | StrOutputParser()
    reflection = chain.invoke({"tweet": state['tweet']})
    state['reflection'] = reflection
    return state
```

The function takes an AgentState object as input, which holds the current state of the agent, including the tweet to be analyzed. Inside the function, a chain is created using the reflection_template, an LLM, and a StrOutputParser. This chain processes the tweet by invoking the reflection template and parsing the output to generate a reflection. The resulting reflection is then stored back into the state object under the key reflection, allowing the agent to use this reflection in subsequent steps. This setup is integral to building a modular and reusable workflow that systematically improves tweet content based on structured feedback.

There is the function to improve the tweet:

```
def improve_tweet(state: AgentState) -> AgentState:
    chain = improve_tweet_template | llm | StrOutputParser()
    improved_tweet = chain.invoke({"tweet": state['tweet'],
    "reflection": state['reflection']})
    state['improved_tweet'] = improved_tweet
    return state
```

This also takes in the AgentState object as input. There is also the same approach with the use of the improve_tweet_template, an LLM, and a StrOutputParser as well as the use of state.

We define the graph:

```
workflow = StateGraph(AgentState)
```

We add some nodes for the reflection and the improvement of the tweet:

```
workflow.add_node("reflect", reflect)
workflow.add_node("improve_tweet", improve_tweet)
```

The same goes for the edges of the graph:

```
workflow.add_edge("reflect", "improve_tweet")
```

Then we set an entry point for the graph:

```
workflow.set_entry_point("reflect")
```

We compile the graph:

```
graph = workflow.compile()
```

We create a function to run the graph:

```
def improve_tweet_with_reflection(tweet: str) -> str:
    result = graph.invoke({"tweet": tweet})
    return result['improved_tweet']
```

We invoke the graph with the original tweet passed as input. The workflow processes the tweet through various nodes, including generating a reflection and then improving the tweet based on that reflection. The final output, an enhanced version of the tweet, is extracted from the result and returned as the function's output.

Then we use an example of this agent:

```
if __name__ == "__main__":
    original_tweet = "I think AI is cool and will change
    everything."
```

```
improved_tweet = improve_tweet_with_
reflection(original_tweet)
print(f"Original tweet: {original_tweet}")
print(f"Improved tweet: {improved_tweet}")
```

The original_tweet is defined with the text "I think AI is cool and will change everything." The function improve_tweet_with_reflection is then called with this tweet, and the resulting improved_tweet is stored. Finally, both the original and improved tweets are printed to the console, allowing you to see the before-and-after results of the tweet enhancement process.

In fact, we can create a visualization of this agent:

```
from IPython.display import Image, display

try:
    display(Image(graph.get_graph().draw_mermaid_png()))
except Exception:
    # This requires some extra dependencies and is optional
    pass
```

The IPython.display module generates a graphic representation of the workflow using Mermaid diagrams, rendered as a PNG image. The graph.get_graph().draw_mermaid_png() function is called to create this visual, and display(Image(...)) is used to show it within the environment. If an error occurs, such as missing dependencies required for rendering the image, the except block catches the exception and passes, allowing the script to continue running without interruption.

Figure 9-1 shows what it looks like.

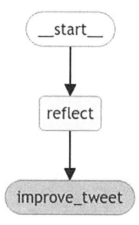

Figure 9-1. *A visualization of a reflection agent in LangGraph*

Persistence

Persistence is a critical feature in AI workflows, especially when it comes to applications that need to maintain context across multiple interactions. In the context of LangGraph, persistence refers to the ability to store and retrieve the state of a graph after its nodes have been executed. This means that an AI agent can "remember" what it has done in the past, allowing it to pick up where it left off after an interruption or between different sessions. This is particularly useful for applications where user input plays a significant role, as the state of the agent can be saved and resumed without losing any progress.

Using a feature called checkpointer, LangGraph provides a streamlined way to store this state data in various persistent storage systems, such as SQLite, Postgres, or MongoDB. When an agent pauses—whether waiting for user input or due to another event—its state is saved and can be retrieved later to continue processing from where it left off. This ability to store state over time also enhances debugging, tracking history, and supporting multiple user sessions, making persistence essential for robust, production-grade AI applications. With LangGraph's checkpointers, you can ensure that no matter what happens during an interaction, your application will be able to pick up right where it left off, seamlessly and efficiently.

We'll write a program that illustrates the use of persistence. The scenario is for a travel booking assistant.

For the program, we have this setup:

```
from langgraph.graph.message import add_messages
```

add_messages, imported from langgraph.graph.message, is a function used to append or combine messages, so as to handle the concatenation of chat messages in a stateful way as the graph processes interactions.

Then we create different tools:

```
from langchain_core.tools import tool

@tool
def book_flight(destination: str):
    """Book a flight to the specified destination."""
    return {"confirmation": "FL12345", "destination": destination}

@tool
def book_hotel(location: str):
    """Book a hotel in the specified location."""
    return {"confirmation": "HT98765", "location": location}

@tool
def book_car_rental(location: str):
    """Book a car rental in the specified location."""
    return {"confirmation": "CR56789", "location": location}

tools = [book_flight, book_hotel, book_car_rental]
```

The tool decorator from langchain_core.tools is used to define functions that can be invoked by an AI agent as part of a workflow, turning regular Python functions into callable "tools" within a LangChain or

LangGraph environment. In this case, three distinct tools are defined for booking travel services: book_flight, book_hotel, and book_car_rental. Each function takes a specific input—destination for flights and location for hotels and car rentals—and returns a confirmation number along with the requested information. These tools simulate actions that an AI agent can perform.

Then we set up the OpenAI model:

```
from langchain_openai import ChatOpenAI

model = ChatOpenAI(temperature=0, streaming=True)
bound_model = model.bind_tools(tools)
```

We create two functions for the workflow:

```
from typing import Literal
def should_continue(state: TravelState) -> Literal["action",
"__end__"]:
    last_message = state["messages"][-1]
    if not last_message.tool_calls:
        return "__end__"
    return "action"

def call_model(state: TravelState):
    response = model.invoke(state["messages"])
    return {"messages": response}
```

The Literal import from typing is used to specify that a function's return value must be one of a few specific string options, providing more precise control over the flow of logic. In the should_continue function, the agent examines the state—specifically, the last message in the "messages" list—to decide whether to continue with an action or finish the current task. If no tool calls were made in the last message, it returns __end__, indicating that the process is complete. Otherwise, it returns "action," signaling that further steps are needed.

The `call_model` function is responsible for invoking the AI model to generate a response based on the current state (the list of messages). It returns the generated message wrapped in a dictionary, allowing the agent to use this output in the next step of its decision-making process.

We set up the workflows for the agents, tools, and edges:

```python
from langgraph.graph import StateGraph, START
from langgraph.prebuilt import ToolNode

tool_node = ToolNode(tools)
workflow = StateGraph(TravelState)

workflow.add_node("agent", call_model)
workflow.add_node("action", tool_node)

workflow.add_edge(START, "agent")
workflow.add_conditional_edges("agent", should_continue)
workflow.add_edge("action", "agent")
```

The imports from `langgraph.graph` bring in the `StateGraph` class, which is used to define a workflow that maintains the state across different nodes, and the `START` constant, which defines the starting point of the graph. Additionally, `ToolNode` from `langgraph.prebuilt` is a preconfigured node responsible for executing tools (in this case, the travel booking tools).

In this code, `tool_node` is instantiated with the predefined tools, such as flight, hotel, and car rental booking functions, making them executable within the workflow. A `StateGraph` object is created to manage the state of the travel agent (represented by the "TravelState"), and two main nodes are defined: the "agent" node, which invokes the model via the `call_model` function, and the "action" node, which runs the tool when the model decides to take an action. The workflow is set up by adding edges: it starts with the "agent" node, which is linked to the "action" node based on conditions. This structure allows the workflow to cycle between decision-making (agent) and action execution (tools), dynamically progressing based on real-time inputs and tool calls.

We then set up the memory:

```
from langgraph.checkpoint.memory import MemorySaver

memory = MemorySaver()
app = workflow.compile(checkpointer=memory)
```

The MemorySaver class from langgraph.checkpoint.memory is imported to provide in-memory storage for the workflow's state, enabling persistence across interactions. By creating an instance of MemorySaver, called memory, the state of the agent can be stored in memory, ensuring that previous interactions are saved and can be accessed later.

The memory object is then passed as a checkpointer when compiling the workflow using workflow.compile(checkpointer=memory). This step ensures that after each node execution in the graph, the state is saved, allowing the agent to "remember" what has already been processed.

Finally, we create the chat for the human and AI:

```
from langchain_core.messages import HumanMessage

config = {"configurable": {"thread_id": "2"}}
input_message = HumanMessage(content="Hello, I want to book a
flight to New York?")
for event in app.stream({"messages": [input_message]}, config,
stream_mode="values"):
    event["messages"][-1].pretty_print()

input_message = HumanMessage(content="I also want to book a
hotel room.")
for event in app.stream({"messages": [input_message]}, config,
stream_mode="values"):
    event["messages"][-1].pretty_print()
```

The `HumanMessage` class from `langchain_core.messages` is imported to represent user input in the form of human-readable messages.

The config dictionary includes a key "configurable" that specifies a `thread_id`, which ensures that the agent can track the conversation within the same session. In this case, the `thread_id` is set to "2," allowing the agent to maintain context throughout the interaction.

In the first part of the interaction, an instance of `HumanMessage` is created with the content "Hello, I want to book a flight to New York?", and it is passed to the `app.stream` function along with the config. The agent processes this message and returns a response, which is then printed.

The second message, "I also want to book a hotel room.", is handled in a similar way, continuing the conversation in the same thread.

This is the output:

> Human: Hello, I want to book a flight to New York?
>
> AI: Great! When would you like to travel to New York and from which city would you be departing?
>
> Human: I also want to book a hotel room.
>
> AI: Of course! Would you like assistance in finding a hotel in New York as well? If so, what are your preferences for the hotel such as budget, location, amenities, etc.?

So as you can see, in the last thread from the AI, it recognizes the city. In other words, there is persistence across the different nodes and calls to the LLM.

LangSmith

LangSmith is a powerful platform designed to build production-ready LLM applications. It makes it easier for developers to manage the complexities of these systems. With LangSmith, you can keep a close eye on your application's performance and progress, allowing for faster deployment with increased confidence. Its seamless integration with LangChain and LangGraph, along with the ability to operate independently, provides a flexible and reliable framework for handling LLM solutions.

Working with LLMs can be challenging due to their probabilistic nature. That is, a response can be unpredictable, often producing inconsistent results based on natural language prompts.

Key features of LangSmith include tracing and debugging, where it offers detailed logs of application runs, which help developers understand the entire operation and quickly identify any issues. The platform also provides robust tools for evaluation and testing, enabling developers to create datasets of inputs and expected outputs for automated and manual assessments. In production, LangSmith offers real-time monitoring of essential metrics like latency, cost, and user feedback, ensuring any problems are swiftly addressed.

Collaboration is another core focus, with LangSmith providing features for annotation and feedback, allowing developers to share insights and improve debugging and evaluation. Moreover, the platform offers versioning and comparison tools, enabling users to analyze different application versions side by side and track changes over time.

LangSmith offers three pricing plans tailored to different user needs. The Developer plan, free for one user, includes 5,000 free traces per month, with additional traces billed at $0.05 per trace. The Plus plan costs $39 per user per month, offering 10,000 free traces with the same rate for extra traces. Enterprise pricing is customized, providing features like Single Sign-On, deployment options, and dedicated support. All plans support key features like debugging, testing, and monitoring, with additional collaboration and security options for teams.

Let's take a look at LangSmith. You can register for the service at this URL: smith.langchain.com. Figure 9-2 shows the dashboard.

Figure 9-2. *This is the dashboard for LangSmith*

It's divided into different sections, such as Projects, Datasets & Testing, Annotation Queues, and Prompts.

To get an API key, select the Settings icon on the left side of the dashboard. You will then choose Create API Key. You have two options. One is the Personal Access Token, which is for an individual user. Then there is the Service Key. This is for more advanced capabilities.

For our purposes, we'll select the Personal Access Token.

Next, you will use these commands in the terminal:

```
export LANGCHAIN_TRACING_V2=true
export LANGCHAIN_API_KEY=<your-api-key>
```

With this, a connection will be made to LangSmith. So when you go back to the dashboard, you can then use the different functions to track, debug, and monitor the agent.

Assistant-UI

Assistant-UI (https://www.assistant-ui.com/docs) is a React component library for building chatbot-like UIs. It has an integration with

LangGraph Cloud. You can create a new project using this, or you can use it with existing React projects with the various components.

It's simple to implement. First, here's how you create a new project:

```
npx assistant-ui@latest create my-app
cd my-app
```

To add the API key for accessing OpenAI, we create a new .env file to the project with your OpenAI API key:

```
OPENAI_API_KEY="sk-xxxxxxxxxxxxxxxxxxxxxxxxxxxxxxxxxxxxxxxxxxxxxxxx"
```

To run the application, use the following (as you would with a React application):

```
npm run dev
```

This will display a skeleton for the Chat Agent UI, as you can see in Figure 9-3.

Figure 9-3. *The Assistant-UI screen after running the application*

LangGraph Studio

LangGraph Studio (github.com/langchain-ai/langgraph-studio) is a desktop application for prototyping and debugging LangGraph applications. It offers visual ways to interact with, edit, and debug agent workflows. There is also step-by-step execution and human-in-the-loop—all integrated with LangSmith. It is currently available for macOS and Windows (Linux support is coming soon). It requires Docker to set up. Currently, this application is in the beta phase and is free. Figure 9-4 shows the dashboard for this application.

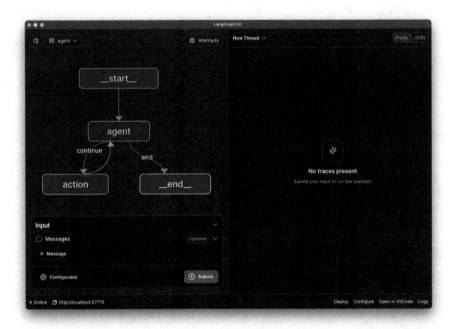

Figure 9-4. *LangGraph Studio screen displaying a LangGraph agent workflow*

Conclusion

LangGraph stands as an effective tool for building stateful, adaptive agents that can handle complex workflows. Its flexibility in creating cyclic structures, persistence, and fine-grained control empowers developers to overcome the limitations of traditional DAG frameworks. By combining LangGraph with LangChain, developers gain access to a comprehensive ecosystem that simplifies the creation of advanced AI systems capable of continuous learning and dynamic interactions.

CHAPTER 10

Haystack

The Haystack framework, developed by Berlin-based startup Deepset, is an open source tool that allows developers to build advanced AI applications using large language models (LLMs). Founded in 2018 by Milos Rusic and Malte Pietsch, Deepset gained early attention for its focus on natural language processing (NLP) solutions, such as training the first German BERT model. Haystack has since evolved into a powerful tool for building custom applications like question answering, semantic search, and Retrieval-Augmented Generation (RAG) systems.

Haystack's strength lies in its modularity and flexibility. This enables developers to integrate components like vector databases, transformer models, and LLMs from platforms such as Hugging Face, OpenAI, or even custom models hosted on various cloud platforms like AWS or Azure. It allows users to connect these components into pipelines, providing full control over the data flow and enabling the creation of robust, scalable AI solution.

Beyond open source tools, Deepset offers an enterprise version, Deepset Cloud, which simplifies the deployment of production-ready NLP applications. Haystack's features make it useful for enterprises looking to build and manage NLP-driven applications with greater ease, thanks to its support for document retrieval, semantic search, and advanced dynamic template generation.

With Deepset continuing to secure significant investment, including $30 million in 2023, the company is poised to further expand its LLM-focused offerings.[1]

Haystack Program

To get a sense of how Haystack works, we'll create a simple program. It will use a Retrieval-Augmented Generation (RAG) pipeline to retrieve and generate answers to a question given a set of documents.

First, we will need to install the framework:

```
pip install haystack-ai
```

Next, we will install few libraries:

```
from haystack import Pipeline, Document
from haystack.utils import Secret
from haystack.document_stores.in_memory import
InMemoryDocumentStore
from haystack.components.retrievers.in_memory import
InMemoryBM25Retriever
from haystack.components.generators import OpenAIGenerator
from haystack.components.builders.answer_builder import
AnswerBuilder
from haystack.components.builders.prompt_builder import
PromptBuilder
```

Here are the explanations:

- Pipeline: Allows you to develop a series of processes ranging from document retrieval to answer generation.

---

[1] https://techcrunch.com/2023/08/09/deepset-secures-30m-to-expand-its-llm-focused-mlops-offerings/

- Document: A single unit or piece of content or text that is used in answering questions.

- Secret: Library for handling private information like API keys for the OpenAI API.

- InMemoryDocumentStore: Library to store the documents that are kept in memory for quick lookup at runtime.

- InMemoryBM25Retriever: Helps to find relevant documents from the document store given the question using BM25 as its method of retrieval. This is a traditional information retrieval algorithm.

- OpenAIGenerator: Library for the OpenAI API.

- AnswerBuilder and PromptBuilder: Classes for instantiating responses and prompts for the LLM.

We instantiate a document in an in-store memory database. We will write three simple documents to it:

```
document_store = InMemoryDocumentStore()
document_store.write_documents([
    Document(content="My name is Jean and I live in Paris."),
    Document(content="My name is Mark and I live in Berlin."),
    Document(content="My name is Giorgio and I live in Rome.")
])
```

These will be the documents from which the pipeline fetches information on certain knowledge.

Then we have the code for the RAG pipeline:

```
# Build a RAG pipeline
prompt_template = """
Given these documents, answer the question.
```

```
Documents:
{% for doc in documents %}
    {{ doc.content }}
{% endfor %}
Question: {{question}}
Answer:
"""

retriever = InMemoryBM25Retriever(document_
store=document_store)
prompt_builder = PromptBuilder(template=prompt_template)
api_key = userdata.get('OPEN_AI_API_KEY')
llm = OpenAIGenerator(api_key=Secret.from_token(api_key))

rag_pipeline = Pipeline()
rag_pipeline.add_component("retriever", retriever)
rag_pipeline.add_component("prompt_builder", prompt_builder)
rag_pipeline.add_component("llm", llm)
rag_pipeline.connect("retriever", "prompt_builder.documents")
rag_pipeline.connect("prompt_builder", "llm")
```

Here's a rundown:

- Prompt Template: A custom template is generated using the PromptBuilder. It takes the retrieved documents as input and creates a prompt for the language model, which asks the model to give an answer.

- Retriever: The InMemoryBM25Retriever is configured to look into the document store for the best matching documents to the query. In this example, it will fetch a document that is very likely to contain an answer.

- PromptBuilder: This constructs the final prompt that goes to the LLM. It inserts the retrieved documents into the template.

- OpenAIGenerator: This module is responsible for generating responses with the use of OpenAI's GPT-based models. The key of the OpenAI API is securely retrieved through "Secret.from_token()".

We construct a question using this code:

```
# Ask a question
question = "Who lives in Paris?"
results = rag pipeline.run(
    {
        "retriever": {"query": question},
        "prompt_builder": {"question": question},
    }
)
```

The pipeline runs, retrieving relevant documents and generating an answer using OpenAI's GPT model.

We print the response:

```
print(results["llm"]["replies"])
```

In this example, the system will likely find the document "My name is Jean, and I live in Paris" for this question's answer.

This setup illustrates how RAG pipelines combine traditional search with the generative power of large language models to answer questions based on documents provided with the help of the Haystack framework.

Haystack Agent with Function Calling

The OpenAIFunctionCaller class of the haystack-experimental package helps to connect models like GPT-4 easier in Jupyter or Colab notebooks. It also supports function calling. That is, it will recognize when to invoke certain functions, say to look something up or make a calculation.

This is a handy component in that it makes the integration of agents into the workflow of Haystack easier. It also handles common problems like API errors or poor connections. This component can even deal with retries for failed attempts. For the most part, this component helps to create more interactive and smart agents, since the AI can decide when to perform particular actions based on the user's input.

Let's take a look at the example. We will create a program that uses a RAG pipeline. For this, the user will ask questions, and the AI agent will dynamically decide to call on various functions like asking about the weather or who lives where.

First, we need to install Gradio, which allows for creating simple web interfaces:

```
pip install haystack-ai gradio
```

Next, we import various frameworks for Haystack that will allow for using memory, RAG, prompts, OpenAI LLMs, and function calling:

```
from haystack.utils import Secret
from haystack.document_stores.in_memory import
InMemoryDocumentStore
from haystack.components.retrievers.in_memory import
InMemoryBM25Retriever
from haystack.components.builders.answer_builder import
AnswerBuilder
from haystack.components.builders.prompt_builder import
PromptBuilder
```

```python
from haystack import component, Pipeline, Document
from haystack.components.builders import PromptBuilder
from haystack.components.generators import OpenAIGenerator
from haystack.components.generators.chat.openai import
OpenAIChatGenerator
from haystack.dataclasses import ChatMessage
from haystack.components.joiners import BranchJoiner
from haystack_experimental.components.tools import
OpenAIFunctionCaller
```

Then we will create the data structure for the weather_fetch function since we will not be using an API (although, in a production program, you would do so):

```python
WEATHER_INFO = {
    "Berlin": {"weather": "mostly sunny", "temperature": 7,
    "unit": "celsius"},
    "Paris": {"weather": "mostly cloudy", "temperature": 8,
    "unit": "celsius"},
    "Rome": {"weather": "sunny", "temperature": 14, "unit":
    "celsius"},
    "Madrid": {"weather": "sunny", "temperature": 10, "unit":
    "celsius"},
    "London": {"weather": "cloudy", "temperature": 9, "unit":
    "celsius"},
}

def get_current_weather(location: str):
    if location in WEATHER_INFO:
        return WEATHER_INFO[location]
    else:
        return {"weather": "sunny", "temperature": 70, "unit":
        "fahrenheit"}
```

243

Next, we define tools for the RAG pipeline and fetching the weather:

```
tools = [
    {
        "type": "function",
        "function": {
            "name": "rag_pipeline_func",
            "description": "Get information about where
            people live",
            "parameters": {
                "type": "object",
                "properties": {
                    "query": {
                        "type": "string",
                        "description": "The query to use in
                        the search. Infer this from the user's
                        message. It should be a question or a
                        statement",
                    }
                },
                "required": ["query"],
            },
        },
    },
    {
        "type": "function",
        "function": {
            "name": "get_current_weather",
            "description": "Get the current weather",
            "parameters": {
                "type": "object",
                "properties": {
```

```
            "location": {"type": "string",
            "description": "The city"}
        },
        "required": ["location"],
    },
  },
 },
]
```

Then we use the tools in the chat_agent:

```
message_collector = BranchJoiner(List[ChatMessage])
chat_generator  = OpenAIChatGenerator(api_key=Secret.
from_token(api_key),model="gpt-3.5-turbo", generation_
kwargs={'tools': tools})
function_caller = OpenAIFunctionCaller(available_
functions={"rag_pipeline_func": rag_pipeline,

                                        "get_current_
                                        weather":
                                        get_current_
                                        weather})

chat_agent = Pipeline()
chat_agent.add_component("message_collector", message_
collector)
chat_agent.add_component("generator", chat_generator)
chat_agent.add_component("function_caller", function_caller)

chat_agent.connect("message_collector", "generator.messages")
chat_agent.connect("generator", "function_caller")
chat_agent.connect("function_caller.function_replies",
"message_collector")
chat_agent.show()
```

The function chat first takes what the user typed and appends it in a list that contains the record of the conversation. This is important as the system needs to know what has been discussed so that answers come out correctly in the context. The chat_agent.run() method sends the user query to an endpoint on OpenAI's API, which then decides if a function call is needed or if a simple text reply will be enough. So, with a query dealing with weather, the system will invoke a pre-set function to fetch current weather. These are then appended to the conversation history for continuity in the chat, and the first response of the assistant is returned and displayed to the user. Figure 10-1 shows the workflow.

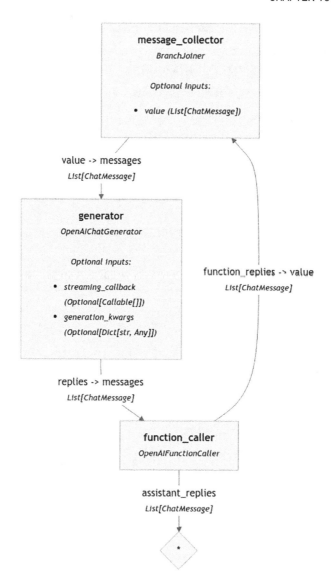

Figure 10-1. *Haystack chat agent architecture*

We will now make an interactive chat interface using Gradio using the
gr.ChatInterface() function. We provide a few example questions that
the user can try, such as "How is the weather in Madrid?" and "Who lives
in London?" These examples give users an idea of how to interact with the
system and some of the things it can do.

```
def chat(message, history):
    messages.append(ChatMessage.from_user(message))
    response = chat_agent.run({"message_collector": {"value":
            messages}})
    messages.extend(response['function_caller']['assistant_
    replies'])
    return response['function_caller']['assistant_replies']
    [0].content

demo = gr.ChatInterface(
    fn=chat,
    examples=[
        "Can you tell me where Giorgio lives?",
        "What's the weather like in Madrid?",
        "Who lives in London?",
        "What's the weather like where Mark lives?",
    ],
    title="Ask me about weather or where people live!",
)
demo.launch(share=True)
```

Finally, we start the interface using the following command: demo.
launch(share=True). It opens a web-based chat interface for inputting
the queries. The argument "share=True" will provide the notebook with a
public URL to use it.

The whole process works once the user types something in the dialog interface. It gets processed through the chat agent that makes a decision on whether to give a text-based answer or make a function call in order to fetch data, such as weather information. The agent then gives a real-time response, and at every new exchange, the conversation history updates, therefore making the system more interactive.

Conclusion

Haystack provides a flexible, modular platform for building sophisticated AI applications, particularly those leveraging RAG pipelines. Its open source framework allows developers to integrate a wide variety of tools, databases, and LLMs, making it ideal for creating not only advanced NLP solutions but also powerful AI agents. These agents can dynamically interact with users, retrieve relevant information, and perform tasks like function calling in real time. Whether through its open source framework or the enterprise-focused Deepset Cloud, Haystack streamlines the development and deployment of AI-driven applications, offering businesses a robust solution for managing and scaling NLP workflows and AI agent interactions.

CHAPTER 11

Takeaways

As we write the last chapter of this book, the category for AI agents has accelerated. The pace of change is really breathtaking.

A major validation of the space is the adoption of this technology from some of the world's largest technology companies. They understand the transformative nature of this technology—how it will go well beyond the typical chatbot approach for generative AI.

Consider the following developments:

- Salesforce's Agentforce: This is a platform that provides conversational capabilities and autonomous agents for many tasks, such as CRM (customer relationship management), marketing, and data management. Agentforce has advanced embedding models that allow for working with complex business workflows and processes. They are also multimodal, handling images, audio, and video. In fact, according to Saleforce's CEO, Marc Benioff, there will be one billion AI agents by the end of fiscal year 2026.[1]

- ServiceNow's Xanadu: This agentic system automates complex processes for customer service management (CSM) and IT service management (ITSM). For example, if a customer reports a Wi-Fi issue, an AI agent can verify

---

[1] https://finance.yahoo.com/news/salesforce-co-founder-and-ceo-marc-benioff-autonomous-ai-agents-will-beat-copilots-155044728.html

network stability, analyze similar past cases, request router details from the customer, and guide the human agent through the next steps—all while following company policies.[2]

- Workday: The company has released various HR and financial management agents. They are based on the training of models for 800 billion business transactions. Among the agents, there is Recruiter. It automates the workflows for identifying talent, outreach, and scheduling interviews.[3]

- Oracle: The company has created more than 50 role-based AI agents for its Cloud Fusion Applications Suite. They span use cases for ERP (enterprise resource planning), HCM (human capital management), SCM (supply chain management), and CX (customer experience).[4]

All these examples point to the strategic importance of AI agents. They also highlight the many opportunities for developers.

According to Jensen Huang, who is the CEO and cofounder of Nvidia, "This is an extraordinary time. In no time in history has technology moved faster than Moore's Law. We're moving way faster than Moore's Law, reasonably Moore's Law squared."[5]

---

[2] https://www.crn.com/news/ai/2024/servicenow-partner-summit-xanadu-release-the-biggest-ai-news

[3] https://www.cio.com/article/3526668/will-workdays-new-ai-agents-set-it-apart-from-competitors.html

[4] https://finance.yahoo.com/news/oracle-adds-powerful-ai-capabilities-132000823.html

[5] https://venturebeat.com/ai/why-jensen-huang-and-marc-benioff-see-gigantic-opportunity-for-agentic-ai/

He also is a big believer in AI agents. He has said that they are a "gigantic" opportunity and are in the "flywheel zone."

Or consider this from Juan Jose Lopez Murphy, who is the Head of Data Science and AI at Globant:

> *AI Agents will have important use cases throughout different industries. The more mature they are in their digital journeys, the more they'll act like software companies. In media and entertainment, they'll be used to process and aggregate even more data for analysis and share faster, more exact personalized recommendations for each user. In healthcare, AI Agents could be used to create even stronger predictions of future diagnoses based on initial symptoms. Agents could also proactively search and test compounds, proteins, new drug developments and many exploration tasks that are very time consuming and that combine a requirement of extensive knowledge and high creativity.*[6]

These are certainly exciting times. And yes, our book is a way to get a starting-off point to participate in this industry. You can then build on this—focusing on those areas you find most interesting.

For the last chapter of this book, we'll provide some takeaways and observations. True, the category for AI agents is moving fast. But for the most part, it seems very clear it's an area that is poised for long-term growth.

Rethinking Software

In this book, we saw how AI agents will be transformative across industries and personal applications alike. From revolutionizing business workflows to enhancing customer experiences, AI agents represent a fundamental shift in how we interact with technology. The potential of these systems

---

[6] This quote is from an interview with the authors of this book.

lies not only in their ability to automate complex tasks but also in their capacity to operate autonomously, learning and adapting in real time. This evolution is already prompting significant changes, especially in the areas of user interface (UI) and user experience (UX) design, which must be rethought to accommodate the unique challenges of AI-driven interactions.

Unlike traditional software that follows predictable, rule-based workflows, AI agents require dynamic interfaces that balance autonomy with user control. Users need transparency into the agent's decision-making process and should be able to intervene when necessary while also benefiting from a seamless, automated experience. This calls for interfaces that can manage complex tasks in the background, yet provide timely updates and options for human input. These changes will fundamentally reshape how software is designed, emphasizing real-time feedback loops, decision logs, and adaptable interfaces that evolve alongside the AI's capabilities.

Moreover, the development process is changing. Whereas conventional development follows a deterministic, step-by-step approach, working with agentic AI relies on probabilistic models that generate outcomes based on a range of potential inputs. This introduces unpredictability, requiring extensive testing, fine-tuning, and ongoing iteration to ensure reliability. Developers must now consider variables like model accuracy, data integrity, and the potential for unexpected outputs. These complexities demand new tools, workflows, and skill sets, as the development process is less about writing rigid code and more about shaping models and algorithms that can adapt to various scenarios.

Moreover, the distinction between traditional automation tools like robotic process automation (RPA) and AI agents is also fading. AI agents are evolving to perform not just repetitive, rule-based tasks but also complex decision-making processes, often without direct human oversight. This shift is driving a rethinking of business models, as the focus moves from subscription-based services to outcome-based pricing, where companies pay for measurable improvements in productivity or cost savings. In this context, AI agents are poised to become a cornerstone of enterprise systems, seamlessly integrating with existing infrastructures and driving efficiency in unprecedented ways.

The Challenges

While AI agents hold tremendous promise, they also come with notable downsides that must be addressed for their widespread adoption. One of the key challenges lies in the frameworks used to build these agents, which are still in their nascent stages. Despite the excitement surrounding platforms like LangChain, AutoGen, and LangGraph, their complexity and evolving nature can make them difficult to work with. Developers often face difficulties orchestrating multiple components—such as memory, tools, and multi-agent systems—into a cohesive workflow. These challenges are compounded by the need for human oversight, as fully autonomous systems are not yet reliable enough to operate independently.

Data management presents another significant obstacle. AI agents rely on vast amounts of data to function effectively, but managing this data—especially in real time—can be overwhelming. Data orchestration, which involves coordinating data from different sources, presents its own difficulties, particularly when systems require frequent updates or process dynamic data types.

Furthermore, these agents often deal with highly sensitive information, raising serious concerns about privacy, security, and governance. Without robust measures in place, the risks of data breaches, unauthorized access, or unintentional misuse increase, potentially eroding trust in the technology. Ensuring that AI agents are secure, private, and adhere to governance standards is crucial for their long-term viability, particularly in sectors like healthcare, finance, and enterprise applications.

Another major limitation of current AI agents is the reliance on transformer models, which have inherent weaknesses. These models, while revolutionary, are prone to hallucinations—where the AI generates false or misleading information—and have cut-off dates for their training data, making them unable to provide up-to-date or entirely accurate information. While progress has been made in improving the reliability of large language models (LLMs), there is still much work to be done to make them more dependable in high-stakes environments. Additionally, transformers are resource-intensive, requiring significant computational power and energy, making them costly and less sustainable at scale.

AI Agent Frameworks

In this book, we have covered several of the leading AI agent frameworks, such as LangGraph, AutoGen, CrewAI, LangChain, and Haystack.

So which one to use? There are no clear-cut answers. Part of this is due to the fact that the industry is moving so quickly.

Yet there are still some general factors to keep in mind. For example, LangGraph approaches agents by using graphs for the decision-making processes. This allows for more granular control of the workflows.

LangGraph will work fine in projects where complex decision-making processes are required, such as in a customer service system, which will take into consideration hundreds of different situations. The framework also provides strong traceability.

AutoGen, on the other hand, is well suited for collaboration. In fact, it can handle many functions out of the box. Think of AutoGen as a system where various agents work together like a committee of experts. This can be helpful with scenarios like major problem-solving and in making advanced chatbots.

Next, there is CrewAI. This framework is one of the most intuitive. Yet it is still powerful. CrewAI is designed to mimic how human teams work. That makes it a good choice for projects that require a number of different roles, such as a virtual project management tool or a creative writing assistant that uses different AIs.

As for LangChain, this is like the Swiss Army knife of AI frameworks: very flexible, packed full of tools, and is a good choice if you want to build something very specific or if you need to combine AI with other types of data or systems. It has been around longer than the others, which generally means there are more examples and community support to help you out.

Then there is Haystack. This framework is particularly good with large datasets for RAG (Retrieval-Augmented Generation), with many integrations for databases and deep NLP (natural language processing) models.

Note that all of these frameworks work seamlessly with LLMs. But each does this differently. There are also important differences when it comes to the size of the application and scalability.

No doubt, a critical factor for an agent framework is the ecosystem. How many contributors does it have? How many times has it been downloaded and starred?

LangChain and LangGraph have perhaps the most vibrant ecosystems. But frameworks like AutoGen, CrewAI, and Haystack have been gaining lots of momentum. AutoGen also has the advantage of the backing of Microsoft. This is certainly important for enterprises.

In choosing between frameworks, consider your project's complexity, your team's strengths, the desired project growth, and special needs. If you are in doubt, it's a good idea to try a couple to find out what works best. Remember, there is no one-size-fits-all answer. The best depends on what you're trying to build and who is going to use it.

Table 11-1 provides comparisons among these frameworks.

Table 11-1. *A Comparison Table for AI Agent Frameworks*

Factor	LangGraph	AutoGen	CrewAI	LangChain	Haystack
Complexity	✓✓✓	✓✓	✓	✓✓✓	✓✓
Ease of Use	✓	✓✓	✓✓✓	✓	✓✓
Multi-agent Collaboration	✓✓	✓✓✓	✓✓	✓✓	✓
Visualization of Workflows	✓✓✓	✓	✓✓	✓	✓✓
Customization Factor	✓✓	✓	✓	✓✓✓	✓✓
Community Support	✓	✓✓	✓	✓✓✓	✓
Learning Curve	Steep	Moderate	Gentle	Steep	Moderate
Scalability	✓✓	✓✓✓	✓✓	✓✓✓	✓✓✓

Integration Capabilities	✓✓	✓✓	✓	✓✓✓	✓✓✓
Use Case Examples	1. Advanced diagnostic systems in healthcare 2. Financial modeling with multiple decision points 3. Adaptive e-learning platforms 4. Complex customer support chatbots	1. Collaborative research assistants 2. Multi-expert consulting systems 3. Complex problem-solving platforms 4. Simulations of multi-entity systems (e.g., economic models)	1. AI-driven project management tools 2. Virtual event planning systems 3. Creative writing assistants with specialized roles 4. Business simulations for training	1. Customized industry-specific AI assistants 2. Data analysis pipelines combining LLMs with other data sources 3. Rapid prototyping of various AI applications 4. Content generation systems requiring fine-grained control	1. Large-scale document search and retrieval systems 2. Question-answering applications over large datasets 3. Information extraction from unstructured data 4. Building conversational AI with access to external knowledge

Legend: ✓ = Fair, ✓✓ = Good, ✓✓✓ = Excellent

Conclusion

In this book, we covered quite a bit about the transformative potential of AI agents, from their foundational components and evolving frameworks to the challenges and innovations that lie ahead. Yet, as comprehensive as this discussion has been, it is still a foundation for further learning and exploration.

As you continue on your journey with AI agents, remember that this field is still evolving, and staying curious and adaptive will be key to mastering its complexities. New advancements will emerge and, with them, fresh opportunities to push the boundaries of what AI can achieve. The potential of AI agents is boundless, and we are only at the beginning of this exciting transformation.

So, we wish you the best of luck as you continue to explore, innovate, and make your mark in this rapidly advancing world of AI agents!

Glossary

AGI (Artificial General Intelligence): A concept in AI that refers to a machine's ability to understand, learn, and apply knowledge across a wide range of tasks, similar to human intelligence.

AI Agents: Software systems that operate autonomously, using artificial intelligence to make decisions, perform tasks, and interact with users or other systems without constant human intervention.

Alignment: The ability of a model to produce responses that meet user expectations, ensuring outputs are coherent, contextually appropriate, and aligned with the desired goals. Techniques like Reinforcement Learning from Human Feedback (RLHF) enhance alignment.

AutoGen: An open source framework for building LLM-based applications that feature multiple agents working together, enabling advanced multi-agent conversational AI systems.

AutoGen Studio: A low-code platform within AutoGen that helps developers create generative AI agents, providing tools for building skills, models, agents, and workflows.

AutoGPT: An early generative AI system that aimed to automate complex tasks with minimal human intervention, known for its initial excitement followed by the realization of its limitations.

Autonomy: The ability of AI agents to independently make decisions and execute tasks without human intervention, relying on their ability to learn, adapt, and respond to new situations in real time.

Conditional Edges: In LangGraph, these are edges that use a function to decide the next node(s) to transition to based on the current state of the graph.

T. Taulli and G. Deshmukh, *Building Generative AI Agents*,
https://doi.org/10.1007/979-8-8688-1134-0

ConversableAgent: A specialized agent in AutoGen designed to manage conversations effectively, handling input, processing it using predefined logic, and generating appropriate responses.

Copilots: Specialized AI agents designed for specific applications or domains, assisting with tasks like content creation, data analysis, or decision-making within a particular context.

CrewAI: A system within OpenAI's offerings that allows for the development of sophisticated conversational agents, integrating multiple AI technologies to create complex interactions.

Delimiters: Tools used in prompt engineering to clearly separate different sections of text, enhancing the accuracy and focus of AI models in processing information.

Directed Acyclic Graph (DAG): A conceptual model used in computer science and mathematics, consisting of vertices connected by directed edges with no cycles, often used in task scheduling and data processing.

Embodied Agents: AI systems that interact with the physical world, often used in robotics or simulated environments to perform tasks like navigation, assembly, or interaction with humans.

Fine-Tuning: The process of training a general AI model on a smaller, task-specific dataset to refine its performance for particular applications.

Generative AI: A branch of AI that creates diverse content such as text, images, videos, and music based on prompts, often utilizing large language models (LLMs).

Goal-Based Agents: AI systems that achieve specific objectives by considering future outcomes and planning their actions accordingly, often using search algorithms to find the most efficient path to a goal.

Google Colab: A cloud-based Jupyter Notebook environment offering free access to GPUs and TPUs, enabling AI practitioners to train and deploy models without local setup.

Gradio: A web-based interface builder that allows developers to quickly create demos for AI models, enabling real-time user interaction with machine learning applications.

Haystack: A framework developed by Deepset, an open source tool that allows developers to build advanced AI applications using large language models (LLMs). It supports applications like question answering, semantic search, and Retrieval-Augmented Generation (RAG) systems.

Hierarchical Agents: AI systems organized in a tiered structure where high-level agents set overarching goals and lower-level agents handle specific tasks to achieve those goals, optimizing efficiency and decision-making.

Hugging Face: A platform and open source community that offers pretrained AI models, datasets, and tools to simplify natural language processing and generative AI applications.

Jupyter Notebook: A web-based interactive development environment that combines live code, visualizations, and text, commonly used for exploratory data analysis and AI model development.

Jupyter Widgets: Interactive elements such as sliders and buttons that can be embedded within Jupyter Notebooks to allow dynamic manipulation of parameters in AI models.

LangChain: A development framework for building generative AI agents that integrate LLMs with various data sources and tools, allowing for the creation of sophisticated AI-driven applications.

LangGraph: An open source framework by LangChain that allows developers to create stateful and multi-actor AI applications, with a focus on dynamic and adaptive agent architectures.

Learning Agents: AI systems that improve their performance over time by learning from experiences, typically using machine learning techniques to refine their actions and decisions.

LLM (Large Language Model): A type of AI model trained on vast amounts of text data to generate humanlike responses, understand context, and perform a wide range of language-related tasks.

Memory: In AI, the ability to retain and utilize information from previous interactions, enabling the system to maintain context, learn from experiences, and provide coherent and personalized responses.

Model-Based Reflex Agents: AI systems that enhance decision-making by incorporating internal models of the environment, allowing them to predict outcomes and make more informed actions.

Multimodal LLMs: Advanced language models that can process and generate information across various data types such as text, images, audio, and video, enabling more versatile interactions.

Multi-agent Collaboration: The interaction of multiple AI agents, each specializing in different tasks, working together to achieve a common goal, akin to how human teams operate.

Natural Language Processing (NLP): A field of artificial intelligence that focuses on the interaction between computers and humans using natural language. It enables machines to understand, interpret, and generate human language.

Nodes: In LangGraph, these are Python functions that encode the logic of agents, taking the current state as input, performing computations, and returning an updated state.

Ollama: A tool that facilitates running large language models locally on personal devices, providing the ability to load and interact with AI models such as Llama 2 and Mistral.

Open Source LLMs and SLMs: Language models developed and distributed openly, allowing for transparency, community collaboration, and customization. Small language models (SLMs) are more efficient, requiring less computational power and tailored for specific tasks.

Outcome-Based Pricing Model: A business model where charges for software or services are based on measurable improvements in productivity, cost savings, or decision-making effectiveness, rather than the number of users or licenses.

Planning: The process by which AI agents determine a sequence of steps to achieve a specific goal, breaking down complex tasks into manageable actions.

Pretrained Models: LLMs that are trained on extensive datasets before being fine-tuned for specific tasks, allowing for efficient adaptation to various applications.

Prompt Engineering: The art and science of crafting inputs that guide AI systems to generate accurate and relevant responses, often involving iterative refinement of prompts.

Proprietary LLMs: Advanced AI systems owned and controlled by private organizations, offering high performance but often with limitations like customization and data privacy concerns.

Reflection Agent: A specialized agent in LangGraph designed to analyze and evaluate its own decisions and actions, enabling it to improve performance over time through self assessment.

Retrieval-Augmented Generation (RAG): An AI technique where external data sources are incorporated during text generation, improving the accuracy of responses by retrieving relevant information.

RPA (Robotic Process Automation): Technology that automates repetitive, rule-based tasks typically performed by humans, often used in conjunction with AI to handle more complex processes.

Semantic Search: A search technique that improves upon traditional keyword matching by understanding the meanings of words and phrases in their broader context, providing more relevant search results.

Simple Reflex Agents: The most basic type of AI agents, operating based on predefined rules that dictate how they should respond to specific sensory inputs without using memory or learning from past experiences.

State in LangGraph: The current snapshot of an application managed by LangGraph, which includes the schema and reducer functions that dictate how updates are applied during workflow execution.

Streaming Outputs: A feature in LangGraph that allows real-time streaming of results from nodes, enabling responsive interactions and immediate feedback during AI operations.

Streamlit: A Python-based tool for creating simple, interactive web applications, useful for visualizing AI model outputs like text or image generation.

Synthetic Data: Artificially created data that mimics real-world data, used to train AI models when real data is scarce or expensive to obtain.

Test-Time Training (TTT): An emerging AI model that processes more data than transformers while consuming less energy, offering a constant model size regardless of the data volume.

Tools: External APIs or software that AI agents use to extend their capabilities, allowing them to perform complex tasks beyond their core functions.

Transfer Learning: A machine learning technique where a model trained on one task is repurposed for a different but related task, leveraging the knowledge gained during initial training.

Transformer Models: A revolutionary architecture in natural language processing that uses attention mechanisms for efficient data processing, outperforming previous models like RNNs.

TPTU (Task Planning and Tool Usage) Framework: A system that evaluates how effectively LLMs can plan tasks and utilize tools, allowing AI agents to dynamically adjust their actions based on ongoing feedback.

Vector Databases: Specialized databases used to store and query vector representations of data, particularly useful in AI applications like semantic search, where similarities between data points are computed based on their vector representations.

Index

A

Accountability, 13, 39
AgentAction, 198
AgentExecutor, 199
AgentFinish, 198
Agentforce, 251
AgentState, 222
AI agents, 2, 3, 41, 46, 81, 209,
    255, 261
  AI development, 101
  autonomy, 11, 12
  comparison table, 257–260
  cost, 15
  definition, 4
  development, 92
  history, 17–19
  memory, 7–9
  planning, 9, 10
  reflection, 5, 6
  tool, 6, 7
  traditional software
    development, 14
  transformer models, 256
  user experience (UX), 12–14
  user interface (UI), 12–14
AI agents code and experiment, 81

AI application deployment, 211
Alignment, 261
allow_delegation=False
    parameter, 116
Ally.ai, 181
AnswerBuilder, 239
approve_or_reject_leave
    function, 159
Artificial General
    Intelligence (AGI), 2, 3, 261
Artificial intelligence (AI), 47, 90
Asana, 24, 25
Assistants API, 66–68, 72–76
Assistant-UI, 232–234
async_execution, 108
Autoencoding language models, 30
AutoGen, 18, 81, 255, 261
  ConversableAgent, 148–150
  description, 147
  group chat, 162–165
  multi-agentic workflow, 148
  reflection agent, 150–157
  Retrieval-Augmented
    Generation (RAG), 167–170
  studio, 171–177
  tools, 157–162
  web search agent, 165–167

© Tom Taulli, Gaurav Deshmukh 2025
T. Taulli and G. Deshmukh, *Building Generative AI Agents*,
https://doi.org/10.1007/979-8-8688-1134-0

GPSR Compliance
The European Union's (EU) General Product Safety Regulation (GPSR) is a set
of rules that requires consumer products to be safe and our obligations to
ensure this.

If you have any concerns about our products, you can contact us on

ProductSafety@springernature.com

In case Publisher is established outside the EU, the EU authorized
representative is:

Springer Nature Customer Service Center GmbH
Europaplatz 3
69115 Heidelberg, Germany